NEW WCDP

新文京開發出版股份有限公司

新世紀．新視野．新文京—精選教科書．考試用書．專業參考書

Modern Business English
Writing Communication

第2版
SECOND EDITION

現代商用英文
書信與應用

廖敏齡 陳永文・編著

2版序 PREFACE

台灣多年來國際貿易蓬勃發展，為台灣賺取大量外匯，對台灣的經濟做了重大的貢獻。而國際貿易在整個溝通的過程中，大部分仰賴英文書信的往返。因此要成為一位優秀的業務人員，除了英文程度要不錯之外，最重要的是溝通的能力。能寫好一封英文商業書信將有助於彼此的了解，也增加交易達成的成功率。

商業書信是公司沉默的業務員，而且通常代表公司對外的重要聯絡方式，所以在撰寫商業書信時，不僅要讓收信人對公司產生良好印象，也要讓人相信寫信者本身就具備工作的熱誠與效率。生意要做成要經過很多的商業活動與流程，每一封的商業書信都是根據不同的階段，依據某種目的而撰寫的。在國際商務上越來越多的人同意，良好的書寫能力是一項必要的商業技能，能讓你在國際商務上擁有強大的優勢。故而要寫好一封溝通良好的書信就需要模擬不同的情境勤加練習。只有透過不斷的練習才能寫出一封好的商業書信。

筆者於企業界從事國際貿易多年，鑒於在網路科技的時代，全球企業仰賴電子化的應用來增加企業的產能與溝通的效能越來越高，因為不管是內部溝通或是外部溝通都大量使用 E-mail 或是 APP 的應用程式以追求更快的速度與效率。因此本書二版除了修正使用更簡明、扼要的英文書寫方式在實務上的商業溝通之外，對行銷的手法的演進與 E 化方式有概要的介紹。另外在報價的部分，本書區分為主動式的報價與被動式的報價二個構面來說明，這樣讀者可以依不同的情境採取更有效的報價方式。最後，在素有亞太區的矽谷之稱的台灣，為因應外資在台灣的人才需求增加，本書將履歷的介紹著重於美式履歷的說明。

希望本書能讓使用者在商用英文信函的書寫能力上提升，更為精進。

謹識

編者簡介 ABOUT THE AUTHORS

廖敏齡 博士

經歷

- 國立台中科技大學兼任副教授
- 亞洲大學兼任副教授
- 美商應用材料公司行銷經理
- 加拿大阿格斯公司財務顧問
- 美國財務協會會員

陳永文 博士

經歷

- 美國加州執業會計師 (CPA)
- 美國管理會計師 (CGMA)
- 美國會計師協會會員 (Member of AICPA)

目錄 CONTENTS

CONTENTS

CHAPTER 06 還價 (Counter Offer/Negotiation)

CHAPTER 07 樣品與規格 (Sample & Specification)

CHAPTER 08 追蹤 (Follow Up)

CHAPTER 09 訂單 (Purchasing Order)

CHAPTER 10 徵信 (Credit Inquiry)

CONTENTS

CHAPTER 15

代理 (Agency)

CHAPTER 16

其他商業信函 (Other Business Correspondence)

CHAPTER 17

求職信 (Job Hunting)

APPENDIX 

附錄 (Appendix)

CONTENTS

CHAPTER

01 商用英文的應用

1.1 何謂商用英文

商用英文與一般英文最大的不同之處在於一般英文大多針對日常會話或是英文文章的撰寫，而商用英文則是針對商業上的需求。因此商用英文上會有其慣用的語法與商業用語，並且會涉及一些貿易知識。

現代商用英文的溝通模式與以往已有重大變革。以前做生意的模式遠比現在疏遠與正式，買賣雙方的關係拘謹且嚴謹。現代商業的行為模式是以非常輕鬆自在的方式進行，不論是會議上、商展上或是書信上都以更自然、更輕鬆友善的語言方式進行，不再像幾十年前那般刻意的咬文嚼字。

1.2 寫好商用英文的要件

1. 了解閱讀者

要寫好商用英文，你必須先知道：

(1) 誰會閱讀這些文件？

(2) 他們為什麼要閱讀這些文件？

想一想閱讀者在看這份文件時，心理會提出什麼問題？他又想從文件中得到什麼資訊？一位良好的商業溝通者通常會思考到閱讀者的需求，進而提供完整的資訊。想像閱讀者提出的各種問題，並在文件中組織好如何清楚回答這些問題，而且重要的問題需首要回答並詳細說明。因為忙碌的商務人士沒有時間仔細閱讀文件內容，所以務必把重要的資訊以醒目的方式凸顯出來，方便閱讀者可以立即找到資訊。其方式有：

(1) 設定醒目的粗體標題 (bold)

(2) 使用數字編號 (numbered list)

(3) 使用項目符號 (bulleted list)

(4) 以編號或字型大小顯示內容的階層

2. 內容要清楚直接

使用的句子要簡單明瞭，讓讀者一看就懂，並非所有從事商務行為活動的人都是英語佼佼者。亦不要用太過正式的語言，想像你正跟讀者說話，自然地表達，對方就可以感受到你的自然與真誠。通常我們大都會使用主動語氣 (active)，少用被動語氣 (passive)。因為被動語氣使你的說話方式聽起來沉悶並且官僚，主動語氣的說法使你的表達方式更有趣生動。

例如：

（✓） We received your order.（主動）

（×） Your order was received.（被動）

（✓） Our staff developed the new system.（主動）

（×） The new system was developed by our staff.（被動）

3. 禮貌與同理心

我們應儘速回復信件，儘量在收信當天回覆。如果對方所提問題無法於當天立即回覆，亦應於當日簡短回答說明原因，以建立友好關係。信件中應尊重對方的觀點，如果覺得對方的意見不公，應謹慎處理，或委婉表達，切勿引起雙方關係的緊張。因此，商業書信上要避免冒犯與攻擊的語氣，而是使用有禮貌但不失尊嚴的寫法。措詞有禮並不是要用像 your kind consideration（您的關照）這種老套用詞，而是指對收件者展現體諒與同理心，亦是尊重收件者的感覺。

例如：

（✓） Here are our best available rates at New York.

（×） Please find below best available rate at New York.

4. 簡短的段落與句子

曾研究探索人們的閱讀速度和理解程度，結果發現為使可讀性提高，一個段落至多不應超過 65 個字，三至五個句子通常就能涵蓋一個想法。如果段落冗長，讀者就必須有強烈的動機才有閱讀下去的動力。使用的文字也盡可能使用一、兩個音節的字，字的音節越少，收件人閱讀書信的速度就越快。使用簡單的句子比複雜的句子更容易溝通，長句會降低閱讀速度，句子的理想長度應該是短於 15 個字。

例如：

（✓） Please let me know the delegates' names and their flight schedule. They will need to liaise with Martin Lim regarding transportation and hotel arrangements when they arrive.

（×） Kindly revert back with flight schedule upon confirmation and also with names of delegates, and liaise accordingly with our Martin Lim upon arrival at airport for transportation and hotel arrangements.

5. AIDA 和 KISS 原則

(1) AIDA 是指：

- Attention（注意力）

  書信一開始就要能吸引讀者的注意力，例如運用標語 (slogan)、驚嘆語或是問句的方式來抓住讀者的注意力。

- Interest（興趣）

  引起讀者興趣之後，要立刻激發他們探索的興趣，說明一下公司的產品與服務的特殊處，持續保持讀者的興趣。

- Desire（慾望）

  創造市場需求的大餅來自於開發與激起消費者的慾望。因此需把你的產品與服務可以為對方帶來的影響與好處一一說明，激起對方購買的慾望。

- Action（行動）

  敦促讀者採取行動，這樣的行動可能一開始是給予讀者一個清楚的流程，請他按流程行動可能有一個意外的驚喜，或是給一個公司網頁的連結，讓他可以很輕鬆的去瀏覽公司產品與服務的網頁。

(2) KISS：

KISS 就是 keep it as simple as possible，也就是要將書信寫的越簡單扼要越好，並非所有從事國際商務行為的人士皆是英文流利的人士。商用英文作為溝通的工具，其目的在於國際業務的擴展與成長，並非英文的修辭學，所以我等商業從事人員切記不可本末倒置。俗話說「黑貓白貓，會捉老鼠的就是好貓」，有些人英文能力並不強卻可以成為超級業務員，就其原因是溝通、說明的能力好，這種人的書信通常淺顯易懂。

6. 8C 原則

(1) Clearness（清楚）

書信中句子的「清楚」是很重要的原則，每個句子最好只含一個意思，讓讀者易於了解。

(2) Conciseness（簡潔）

商業書信的寫法要力求簡潔扼要，避免用詞累贅。「簡潔」必須與「清楚」的原則配合，但切記不可為了縮短文件的長度而忽略重要資訊。

(3) Correctness（正確）

書信的正確包含文件的格式、句子的文法、標點符號、收件人稱呼、日期、事實與數字等是否都正確。

(4) Concreteness（具體）

具體是指給讀者提供「具體的資料」。不要用比喻或是隱晦的方式，也不要用專業術語，要用具體事實與分析結果。

(5) Completeness（完整）

意思要完整，務必包含足夠的資訊，要考慮對方在閱讀書信時想要獲得的資訊有哪些，以確保文件內容完整。

(6) Courtesy（禮貌）

商業上的禮貌不僅指書信上的書寫方式不可有情緒化 (emotional) 或是粗魯 (rude) 的寫法，更重要是處理商業關係的態度。這裡的「禮貌」指的是表示出願意幫忙的真誠意願。

(7) Considerateness（週到）

週到乃指以對方的考量為主，確實考慮對方的需求，主動為其設想並其提出適時的建議。

(8) Character（特色）

一封書信的特色顯示了寫信人的個性，不管是書寫方式、語氣、用詞皆展現寫信者的特質。創造自己的特色更具獨特性而且辨識度更高。

練習題

選擇題：

(　) 1. Which one is not part of AIDA?

(a) Desire　(b) Attraction　(c) Interest

(　) 2. Which is one of the eight Cs of good writing

(a) Coaching　(b) Compelling　(c) Courtesy

(　) 3. Which sentence is better?

(a) Our next meeting will be held at 9 am tomorrow.

(b) Please be advised that our next meeting will be held at 9 am tomorrow.

(　) 4. What is KISS?

(a) keep it as simple as possible　(b) keep it as special as possible

(c) keep it as speedy as possible

CHAPTER

02 商用英文的工具

目前企業上常用的溝通工具有電子郵件 (e-mail)、書信 (letter)、傳真 (fax)、備忘錄 (memo) 等。而電子郵件是目前在「正式」與「非正式」的商業溝通中最被廣泛應用的工具，書信的應用則多是用於「正式」的溝通工具。

2.1 電子郵件 (e-mail)

電子郵件現在是商業溝通模式中使用最頻繁、最廣泛的工具。E-mail 沒有標準的寫法也比較不拘禮節，是一種很快速的溝通工具，比傳真更方便，而且可附加檔案更方便於公司在傳送規格表、圖紙的完整性。雖然電子郵件比較不拘禮節，但是對於商務溝通的事件來說，在商務書信上還是應該要用正式文字，而不要使用表情符號 (emoticons)。

在寫 e-mail 時有些事情必須注意。由於 e-mail 是在只有 24 行的螢光幕上閱讀，所寫的東西最好儘量可以在一個螢光幕上看完，能不超過一頁最好。要把最重要的東西寫在前面，內容有關文法、句子與標點符號都應按英文語法規定。

1. 電子郵件格式

(1) To：收件者。即對方之 e-mail 地址。發出信件之前應再一次確認收件人地址是否正確，不要誤發了，常用住址可儲存於通訊錄中並加以詳細編輯資料。

(2) From：寄件者最好可以看到寄件人的全名，不要只有名沒有姓，這樣比較專業。商務電子郵件的帳號應該要嚴肅專業，不要用暱稱或是玩笑性質的帳號名稱，例如像是 hellokitty@xxx 或是 teddybear@xxx 等等。一般書寫 e-mail 時，系統會自動帶入寄件者的電子郵件帳號。

(3) Cc：副本抄送 (carbon copy)。指將 e-mail 的副本傳給其他人，將事件告之相關人士。

(4) Bcc：密件副本 (blind carbon copy)。指隱藏收信人郵件地址，表示這一欄的收件人不會被其他收件人知道。但是 Bcc 這一欄的收件人可以看到其他收件人的地址。比如說如果你想寫一封郵件給保險公司，但是也想讓你的律師看到這份資料，那就可以將律師的住址輸入 Bcc 這一欄。

(5) Subject：主旨。即是點出此郵件的主題。主旨的寫法多為片語方式而不是句子模式，且片語應不可太長。主旨的寫法應該 SMART，也就是 Specific（明確）、Meaningful（有意義）、Appropriate（適當）、Relevant（切題）、Thoughtful（體貼）。

(6) Attachment：附件。顯示郵件所附加的檔案。e-mail 可將各種文書檔案、表格檔案、圖片或是影像傳給對方。

(7) Salutation：敬稱。對於初次聯絡或是位階較高的客戶，e-mail 的敬稱應該要使用 Dear 開頭，不可用 Hi, Hello 來開頭。等到雙方關係接近，對方也用較輕鬆隨意的開頭時，我們才可以順應對方使用以拉近雙方的商務關係。

(8) Body：本文。基本上郵件的撰寫的基本綱要來自書信 (letter) 的基本架構。e-mail 的本文多使用齊頭式 (Block Style)，但是撰寫的方式更為簡潔。

(9) Complimentary close：結尾敬語。e-mail 的結尾敬語比較沒有信件的嚴謹，而是用比較輕鬆的語句，常用的有例如：Best regards、Best wishes、Kindest regards、Warmest regards、Regards、With love、As always、Thanks…等。

(10) Signature：簽名。簽名檔 (signature block) 就是一個可以設計在每封電子郵件的底端可以插入的圖文檔（.sig 檔）。簽名檔裡最少應該包含你的全名與職稱，另外也可增加公司名稱、電子郵件、個人專線號碼，甚至是公司的地址、網頁與電話傳真等。

Tips 現代的商業溝通工具發展的越發快速，目前及時溝通還有一些網路軟體被廣大採用，例如：line、skype、wechat、messenger、instagram 等。買賣雙方都可在線上直接溝通，但是還是提醒大家，對於重要的議題（如報價、訂單、交期或是賠償等），最好還是將溝通的訊息放在 e-mail 上，必要時得附件寄給主管或是相關人員，一來這樣資訊才完整，二來可以避免日後的紛爭。

To: jackturner@piorneer.com → (1)
From: lindachen@tbi.com.tw → (2)
Cc: judywong@ttm.com.tw → (3)
Bcc: → (4)
Subject: New Project → (5)
Attachment: → (6)

Dear Mr. Turner → (7)

I am so glad that we have been able to find a solution to this. Good luck with future progress on this project.

I'll be here when you decide how we can help you again. → (8)

Best regards, → (9)

Linda Chen → (10)
Sales Manager

▶ 2.2 書信 (letter)

1. 書信內容

Taiwan Bear International → (1)

3F, No. 11, Park Avenue II
Science-Based Industrial Park
Hsin-Chu 30075, Taiwan
Tel: 886-03-5798888
Fax: 886-035978891

February 19, 20xx → (2)

Jack Turner → (3)
Purchasing Dept.
Pioneer Corporation
3080 Bowers Avenue
Santa Clara, CA 95054
USA

Dear Mr. Turner, → (4)

Subject: YOUR ORDER NUMBER 3321 → (5)

We enclose our invoice number 8864 for the domestic electrical appliances supplied to your order dated November 24.

The goods have been packed in three cases, numbers 51, 52, 53, and sent to you today by air, carriage paid. We hope they will reach you promptly and in good condition. → (6)

If you settle the account within two months we will allow you to deduct from the amount due a special cash discount of 1.5%.

Sincerely yours, → (7)

Linda Chen → (8)

Linda Chen → (9)
Sales Manager → (10)

LC/jw → (11)

Enc. → (12)
C.c. → (13)
P.s. → (14)

(1) Letter Head：信頭。信頭就是公司專用信紙上印有公司名稱與商標的地方。書寫商業文件時，最好使用這種印好信頭的公司專用紙，這樣比使用空白紙要專業。

(2) Date：日期。商務書信的內容常與時間有密切相關，因此一定要寫上日期。日期寫法也有不同的方式，目前企業上多使用美式寫法。

美式寫法（月／日／年）7/12/2021 July 12, 2021

英式寫法（日／月／年）12/7/2021 12th July 2021

(3) Recipient's address (inside address)：收件人名稱及資料。

(4) Salutation：稱謂。務必把收件人的頭銜寫對，不同的專業有不同的頭銜，醫生和博士的頭銜是 Dr.；大學教授可能是 Dr.（博士），也有可能是 Professor（教授）。因此一定要查清楚才不會寫錯。

商務書信上的寫法通常以「頭銜＋姓」的方式稱呼對方，如 Dear Dr. Brown；如果不知頭銜，則男士用 Dear Mr. Brown；女士則用 Dear Ms. Brown（已婚或是未婚通用）。除非對方明確要求你使用 Mrs. Brown（已婚）或是 Miss（小姐）。

如果我們連對方的性別也不知道，那就寫 Ladies and Gentleman 或是 Dear Sir or Madam 或是 To whom it may concern。記得不要把姓跟名都寫出來，例如：Dear John Brown，這樣是不禮貌的寫法。

在稱謂之後的標點符號應該冒號，例如：「Dear Mr. Brown:」。但是目前業界最常使用的是逗號，例如：「Dear Mr. Brown,」。也可以完全不加標點符號，例如：「Dear Mr. Brown」。

(5) Subject：主旨。在信件中的主旨可加可不加。加上這一行主旨是為加強信件的標題，這樣有助於收件人立刻了解信件內的主題。主旨的寫法也有很多種，大部分是用「粗體字」、「顏色字」、「底線」或是「大寫字」來凸顯。

(6) Body：本文。本文為信件最主要的部分，通常由段落組成。本文應該要簡潔扼要，內容要有組織，上下文要連貫，以利收件人得到需要的資料。本文通常可組織成三個段落：

第一段：主旨段。在第一段裡需開宗明義的清楚明確說明這一封信的目的與主旨，因此第一段的段落不會很長，通常為一到三個句子所組成。

第二段：說明段。中間的段落應該來詳細說明本信件的主題，並給予重要的細節與資料的輔助說明。段落應儘量簡短，但是如果要說明的資料太多，或是複雜性很高，就需要分段說明，要注意每一段落只說明一個小主題，才不至於使段落看起來太長不容易閱讀。另外也可用項目符號或是數字編號編成列表式的說明方式。

第三段：結束段。結論的內容應與第一段相呼應。你可以在結束段裡進行感謝、推薦、拒絕、期待未來合作或是表示提供願意協助等總結的想法。因此，結束段也同第一段一樣通常不會超過三個句子所組成的段落。

(7) Complimentary close：結尾敬語。結尾敬語後面需加上逗號，並且只有第一個英文字要大寫。書信常用的結尾敬語有：

- Sincerely yours,
- Faithfully yours,
- Yours truly,

(8) Signature：簽名。「信件」都要簽名以示負責。因此發信人需特別謹慎審閱發出的信。因為這不僅是代表公司的文件，更可能在為來假設有訴訟的情況時，更是成為證據的文件，因為上面有負責的人已簽名。

(9) Name：姓名。這是要打字的全名。在發信人簽名之後需打全名，因為簽名比較個性化，怕無法知曉發信人的全名。

(10)Title：職稱。公司所賦予的頭銜，例如：CEO（總裁）、CFO（財務長）、COO（營運長）、President（董事長／總經理）、VP(Vice President)（副董事長／副總經理）、Director（處長）、Manager（經理）等。

(11)Identification initials：發信人及打字人的姓名縮寫。LC 為發信人（簽名人），通常是老闆。jw 為打字者，通常是祕書。如果發信人與打字者同為一人時，則不需註明。目前實務上的商務書信，大多不會打上此資料。需要打上此一資料大部分是組織很大，公司可能設有祕書處，職務上有主任祕書與數位祕書人員，為區分是哪一位負責的書信就需要打上此資料。

(12)Enclosure/Attachment：附件。可打 Enc. 或是 Encl. 附於信件後的其他文件資料。

(13)Carbon copy：副本抄送。說明此信件將有副本寄給第三人或是相關單位。

(14)Postscript：附註。若信件已經打好，想另外補充幾句或談些與信件內容較不相關的事情時，可在信件的最下方做附註。但是這不是一個好的書信寫法，應避免使用。

2. 書信格式：書信格式有三種，目前最常用為「齊頭式」。

(1) 齊頭式 (full block/block style)：齊頭式是目前商務書信格式中最常用的一種格式。在齊頭式中，每一個部分都從左邊的邊界起頭。不同的段落則以空行的方式隔開（如圖一）。

(2) 改良齊頭式 (modified block style)：改良齊頭式是將 Date、Complimentary close 與 Signature 移至中間往右書寫（如圖二）。

(3) 縮排式 (indented style)：縮排式是比較老式的商業書信格式，目前使用較少。縮排式除了將 Date、Complimentary close 與 Signature 移至中間往右書寫之外，並需在每一個段落的開頭都空 4~8 個空格，關鍵是空格位置前後要一致（如圖三）。段落縮排的可行作法：

- 可以按幾次「空白鍵」
- 可以按一次「tab」鍵
- 可以利用段落格式設定的功能

圖一：齊頭式

Taiwan Bear International
3F, No. 11, Park Avenue II
Science-Based Industrial Park
Hsin-Chu 30075, Taiwan
Tel: 886-03-5798888
Fax: 886-035978891

⇨ 1-3 blank lines

February 19, 20xx

⇨ 1-3 blank lines

Jack Turner
Purchasing Dept.
Pioneer Corporation
3080 Bowers Avenue
Santa Clara, CA 95054
USA

⇨ 1-3 blank lines

Dear Mr. Turner,

⇨ 1 blank line

Subject: YOUR ORDER NUMBER 3321

⇨ 1 blank line

We enclose our invoice number 8864 for the domestic electrical appliances supplied to your order dated November 24.

⇨ 1 blank line

The goods have been packed in three cases, numbers 51, 52, 53, and sent to you today by air, carriage paid. We hope they will reach you promptly and in good condition.

⇨ 1 blank line

If you settle the account within two months we will allow you to deduct from the amount due a special cash discount of 1.5%.

⇨ 1 blank line

Sincerely yours,

⇨ 3-4 blank lines

Linda Chen

Linda Chen
Sales Manager

⇨ 1 blank line

Enc.

圖二：改良齊頭式

Taiwan Bear International
3F, No. 11, Park Avenue II
Science-Based Industrial Park
Hsin-Chu 30075, Taiwan
Tel: 886-03-5798888
Fax: 886-035978891

February 19, 20xx

Jack Turner
Purchasing Dept.
Pioneer Corporation
3080 Bowers Avenue
Santa Clara, CA 95054
USA

Dear Mr. Turner,

Subject: YOUR ORDER NUMBER 3321

We enclose our invoice number 8864 for the domestic electrical appliances supplied to your order dated November 24.

The goods have been packed in three cases, numbers 51, 52, 53, and sent to you today by air, carriage paid. We hope they will reach you promptly and in good condition.

If you settle the account within two months we will allow you to deduct from the amount due a special cash discount of 1.5%.

Sincerely yours,

Linda Chen

Linda Chen
Sales Manager

Enc.

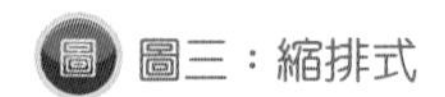
圖三：縮排式

Taiwan Bear International
3F, No. 11, Park Avenue II
Science-Based Industrial Park
Hsin-Chu 30075, Taiwan
Tel: 886-03-5798888
Fax: 886-035978891

February 19, 20xx

Jack Turner
Purchasing Dept.
Pioneer Corporation
3080 Bowers Avenue
Santa Clara, CA 95054
USA

Dear Mr. Turner,

Subject: YOUR ORDER NUMBER 3321

We enclose our invoice number 8864 for the domestic electrical appliances supplied to your order dated November 24.

The goods have been packed in three cases, numbers 51, 52, 53, and sent to you today by air, carriage paid. We hope they will reach you promptly and in good condition.

If you settle the account within two months we will allow you to deduct from the amount due a special cash discount of 1.5%.

Sincerely yours,

Linda Chen

Linda Chen
Sales Manager

Enc.

3. 信封打法

西式信封打法和中式不同，西式信封為橫式打法，中式信封為直式打法。目前很多企業為使中、西式寫法一致都將信封的印製一律用西式的橫式方式。西式信封發信人公司名稱與地址都是在左上角，收件人的公司名稱與地址則是在正中間，郵票貼在右上方。其他註記事項，例如印刷品 (Printed Matter)、快遞 (EXPRESS)、掛號 (REGISTERED) 等，則可打在左或是右下的空白處。

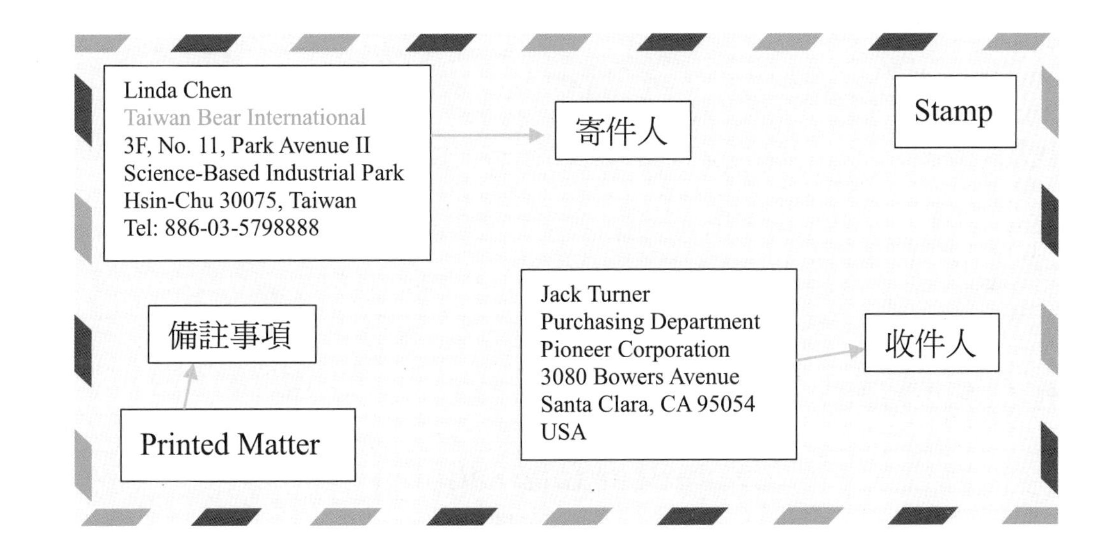

4. 傳真 (fax)

傳真沒有固定的原則或是格式，比較要注意的是要傳真出去的文件應該都要附有一個封面頁 (cover letter)，而傳真文件的撰寫原則就跟商業書信一樣，基本上就是要清楚、簡潔、扼要、禮貌。

範例 1：傳真封面頁

Taiwan Bear International
3F, No. 11, Park Avenue II
Science-Based Industrial Park
Hsin-Chu 30075, Taiwan
Tel: 886-03-5798888
Fax: 886-035978891

FAX

Company:	**Date:**
Attn:	**Pages:**
Fax No.:	

範例 2：傳真封面頁

Taiwan Bear International
3F, No. 11, Park Avenue II
Science-Based Industrial Park
Hsin-Chu 30075, Taiwan
Tel: 886-03-5798888
Fax: 886-035978891

FAX

To:	From:
Company:	Date:
Fax No.	Phone:
Subject:	No. of Pages:

5. 備忘錄 (memo)

備忘錄是辦公室內或是集團內部用來溝通訊息的工具。由於備忘錄屬於內部使用的溝通工具，因此不像一般的商務文件那麼正式。但是備忘錄依舊是用來傳達訊息的，所以在書寫時還是要注意內容需清楚、簡潔、正確。二十世紀之後由於 internet 的發展迅速，目前大部分公司的內部溝通工具已有多元化的 e 化發展，比較少用 memo 的方式了。

範例：

Taiwan Bear International

MEMO

To: Information Design Team
From: John Lin
Cc: Linda Chen
Date: Jan. 11, 2021
Re: Conference travel request

The senior managers reviewed your request to take your team to the "Teambuilding International" conference in Shanghai on August 25. We appreciate your interest in building good working relationships within our office and in our branches worldwide. However, in the current economic situation, our shareholders are unlikely to approve the expense. We are sorry to have to decline your request.

Please feel free to resubmit your request for next year's conference, when we may be in a better position to approve travel.

練習題

選擇題：

(　) 1. Which is not an appropriate way of writing an attention line?

(a) Attention: Ben Lin　(b) Attn. Ben Lin　(c) Attention Ben Lin

(d) all are acceptable.

(　) 2. Which is not an appropriate salutation for a business letter?

(a) Dear Mr. Jorden　(b) Hello　(c) Ladies and Gentlemen

(　) 3. Which is a correct format for the date in a business letter?

(a) 21-09-21　(b) September 21, 2021　(c) 9/21/2021

(　) 4. Which is correct?

(a) Yours Truly　(b) yours truly　(c) Yours truly

問答題：

1. 請標示書信中的欄位名稱。

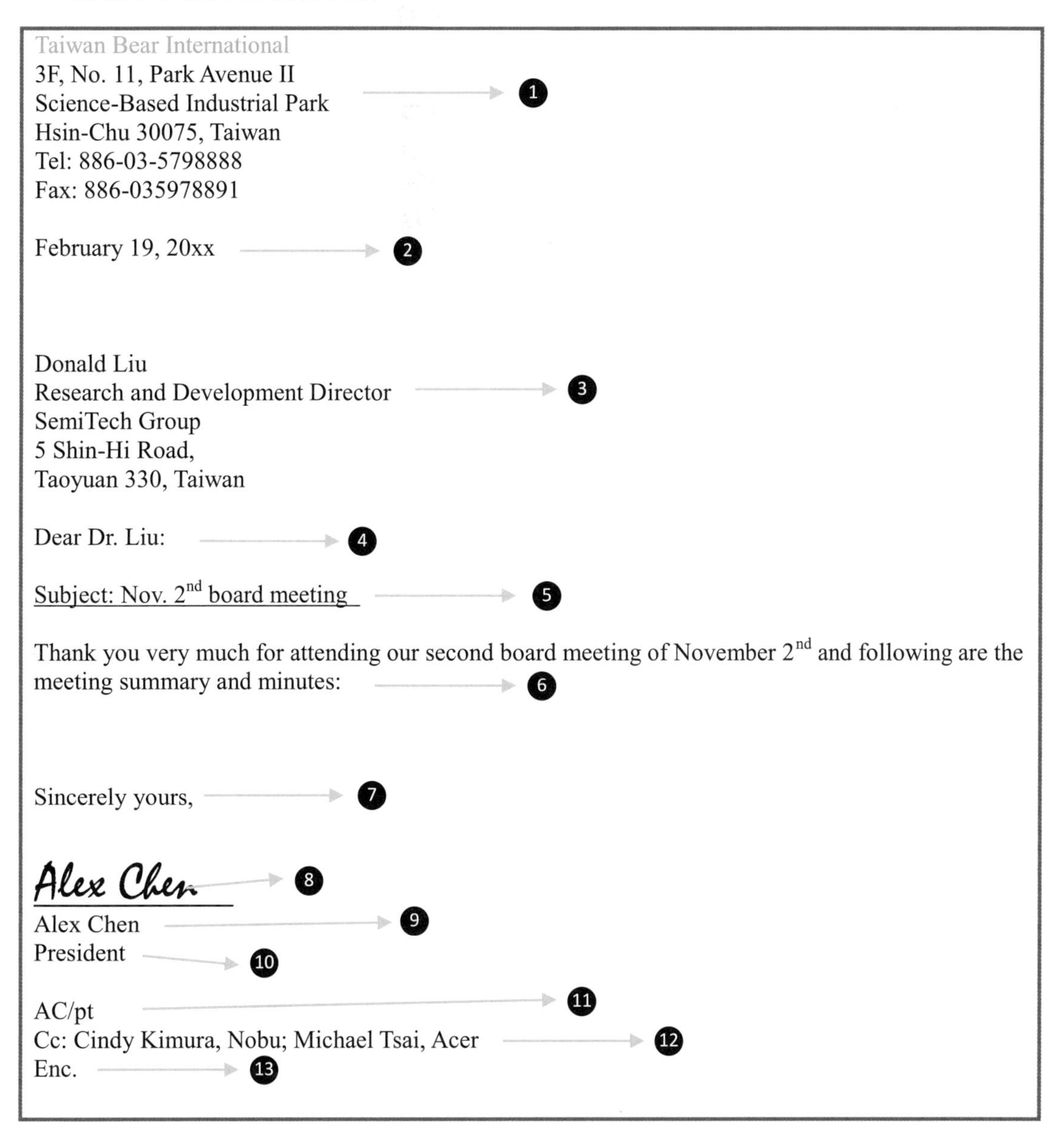
Taiwan Bear International
3F, No. 11, Park Avenue II
Science-Based Industrial Park → ❶
Hsin-Chu 30075, Taiwan
Tel: 886-03-5798888
Fax: 886-035978891

February 19, 20xx → ❷

Donald Liu
Research and Development Director → ❸
SemiTech Group
5 Shin-Hi Road,
Taoyuan 330, Taiwan

Dear Dr. Liu: → ❹

Subject: Nov. 2nd board meeting → ❺

Thank you very much for attending our second board meeting of November 2nd and following are the meeting summary and minutes: → ❻

Sincerely yours, → ❼

Alex Chen → ❽
Alex Chen → ❾
President → ❿

AC/pt → ⓫
Cc: Cindy Kimura, Nobu; Michael Tsai, Acer → ⓬
Enc. → ⓭

2. 請就第一題書寫信封

3. 請就第一題書寫傳真的封面

4. 請就下列情境寫一份 memo：人力資源處長將告訴公司所有員工於今年 11 月底前，所有員工需於電腦問卷上完成「員工滿意度」調查。

CHAPTER

03 促銷信

Promotion/Sales Letter

▶ 3.1 促銷信函的要點

公司成立之後，首要的工作便是尋找客戶，並開始建立商務關係 (business relationship)。在尋找潛在的客戶，我們永遠需要主動出擊，積極透過各種途徑推銷公司產品。如果對方沒有回應，更應該再接再厲的追蹤 (follow up)。

在寫推銷信函時，需注意五個要點：

- 注意 (Attention)：必須以醒目的標題，或是用問題的方式，或是具有挑戰的語氣來吸引對方的注意。
- 興趣 (Interest)：提到會吸引對方的事物，訴諸特殊的動機。
- 渴望 (Desire)：引發對方的渴望而非對方的需求 (needs)，才能開創市場。要強調產品的獨特性、話題性和設計的新穎，並指出我們的產品與服務可以為對方帶來的好處與影響。
- 說服 (Conviction)：在這裡需說服對方，讓他們對我們的產品產生信任感。可以藉由說明公司已在業界創業幾年，或是公司可以提供什麼樣的保證，或是可以提供事實與數據來佐證我們的說法。
- 行動 (Action)：讓對方想要購買，並且進一步了解。可以暗示或是清楚的告訴對方接下來應採取的步驟。我們必須設計一個讓對方可以很容易就採取行動的作法，例如：可以提供回郵信封，或是一個電子回覆的快速連結。

▶ 3.2 促銷的途徑與管道

一個成功的銷貨人員 (sales) 除了對公司產品必須非常了解之外，也必須對產業與競爭者詳加研究，才可以知己知彼，百戰百勝。要成功推銷產品至其他國家，我們必須研究對方的國家法規限制，例如：有沒有管制進口、有沒有違反貿易限制 (antitrust) 的風險、有沒有配額 (quota) 的問題等。另外更要了解對方國家的民俗風情才不會有不當的推銷。

產品推銷的方法與途徑有很多種方式，每一個方式都有不同的效果，一個成功的銷售人員需就不同的客戶與情境挑選最恰當的方式進行，才可以達到事半功倍並節省公司資源的作法。

1. 參展：參展的最大好處是可以直接在展場找到潛在的客戶，並且直接面對，可以當面推銷與討論，並且交換名片，有利於後續追蹤。缺點就是成本太高，而且競爭者也會展出相關或類似的商品。

2. 寄推銷信函：因為網際網路 (internet) 的發達，所以利用電子郵件 (e-mail) 作為推銷的手段成為最經濟的方式，因此最常被使用。但是也因為很多人都收到這樣的推銷信件，如果沒有特別吸引人的商品或是服務，很容易就被丟到垃圾信件匣，所以成功的機率不是很高。

3. 老客戶或朋友引薦：經由現有的客戶或是朋友的推薦，這種方式的成功機率很高，應該要多加把握。

4. 透過政府組織：政府組織有外貿協會的圖書館可以查閱進出口廠商名錄，也可經由貿協的電腦查；或是向商會、工會索取會員名單；或是去函各國貿協請求刊登廣告或是推薦進出口廠商名單。

5. 電腦上搜尋：從網路上搜尋相關的客戶名單。可以訂閱外貿協會的「貿協商情周報」(Trade Weekly)，每週都有最新的資料。

6. 社群媒體上的廣告：透過社群媒體的方式將公司的商品拍成影音模式放上社群網站去吸引潛在的客戶，知名的社群媒體像是 YouTube 或是 facebook 等。

7. 報章、雜誌上的廣告：從各種相關的雜誌、廣告媒體尋找潛在客戶。這是一種比較早期的傳統推銷模式。

8. 找代理：如果對當地的法令或是非英語系作為商務語言的國家，在銷售商品時可能會遇到極大的障礙，此時可以先透過代理的方式來進入這樣的市場。

9. 直接拜訪客戶：這是對已經是公司的舊客戶做定期的拜訪與追蹤，藉以穩定雙方的貿易關係。

範例 3-1　賣方：對潛在客戶通知參展訊息

To: purdept@piorneer.com
From: lindachen@tbi.com.tw
Cc:
Bcc:
Subject: Trade Show in US
Attachment:

Dear Sir,

We obtained your name and address from the Taiwan Trade Center. We are a manufacturer and exporter in Taiwan, specializing in binding machine.

We are pleased to invite you to visit our booth (23321) at the Chicago International Fair from Feb. 10 to Feb. 15, 2021. Visit our booth and you will not be disappointed.

Best regards,

Linda Chen
Sales Manager

翻譯

我們從台灣貿易中心取得您的姓名和地址。 我們是台灣專門從事裝訂機的製造商和出口商。

我們很高興地邀請您參觀我們在 2021 年 2 月 10 日至 2 月 15 日在芝加哥國際商展，我們的位置是 23321。參觀我們的展位，您不會感到失望。

範例 3-2　賣方：對曾經有過貿易的客戶通知參展訊息

To jackturner@piorneer.com

From lindachen@tbi.com.tw

Cc

Bcc

Subject Trade Show in US

Attachment

Dear Mr. Turner

How are you doing recently? I have not received your message since you placed an order to us two years ago. I would like to let you know that our company will have a booth (23321) at the Chicago International Fair from Feb. 10 to Feb. 15, 2021.

I would like to invite you to visit our booth and introduce our new products to you. You will be regret if you miss this excellent opportunity.

Best regards,

Linda Chen
Sales Manager

翻譯

最近怎麼樣？自兩年前您向我們下訂單後，我就沒有收到您的信息。我想告訴您，我們公司將參加 2021 年 2 月 10 日至 2 月 15 日的芝加哥國際商展，我們的攤位編號是 23321。

我想邀請您參觀我們的展場並向您介紹我們的新產品。如果您錯過了這個絕好的機會，你將會感到後悔的。

範例 3-3　賣方：對舊客戶通知參展訊息

To	jackturner@piorneer.com
From	lindachen@tbi.com.tw
Cc	
Bcc	
Subject	Trade Show in US
Attachment	

Dear Jack,

This is a notice that we will participate Chicago International Fair from Feb. 10 to Feb. 15, 2021 and our booth number is 23321.

As usual, our company has continued to develop newer and better products, which have been launched recently, and will be showing them at the Trade Show. We would like to invite you to visit our booth and to have dinner at your available time.

Please let me know when you will visit us and I can arrange the dinner in advance.

Best regards,

Linda Chen
Sales Manager

翻譯

這是一個通知書，通知您我們將於 2021 年 2 月 10 日至 2 月 15 日參加芝加哥國際商展，我們的展位號碼是 23321。

像往常一樣，我們公司不斷地開發並推出更好的的產品，並將在貿易展上展示它們。我們想邀請您參觀我們的展位，並想邀請您方便的時間一起晚餐。

請讓我知道您何時會拜訪我們，我可以提前安排晚餐。

範例 3-4　賣方：公司參展後的後續追蹤客戶

To	jackturner@piorneer.com
From	lindachen@tbi.com.tw
Cc	
Bcc	
Subject	Trade Show of Chicago International Fair
Attachment	

Dear Mr. Turner,

It was a great experience to present you our new products at Chicago International Fair. I have memorized your opinions and suggestions and brought your valued ideas back to our R&D department.

I believe you will be interested how your opinions to improve our products' function. Would you like to receive our samples and quotation?

Best regards,

Linda Chen
Sales Manager

翻譯

在芝加哥國商展上向您介紹我們的新產品是一次非常棒的經驗。我記住了您的意見和建議，並將您的建議帶回到我們的研發部門。

我相信您會對您的意見是如何改善我們產品的功能感興趣。您想收到我們的樣品和報價嗎？

範例 3-5　賣方：專業雜誌上找到潛在客戶

To jackturner@piorneer.com

From lindachen@tbi.com.tw

Cc

Bcc

Subject Silicon Rubber Pads

Attachment

Dear Mr. Turner,

We got your name from CPU magazine and we also know that you are looking for Silicon Rubber Pads. Our company has been working in this field over 20 years and our customers include Texas Instruments, Motorola, Hitach, etc.

Our company has received great honor from Texas Instruments as a "Supplier Excellence Award." In addition, Hitach recognizes us a trustworthy supplier. All of these prove that we not only provide the most competitive prices, best quality and performance, but also have the ability to help customers to solve their problems.

Recently, we have expanded our factory space and installed more up-dated machines, we have the confidence to meet your requirement under our advantageous position. Please visit our web: www.tbi.com.tw for more details. If you are interested in any products, please don't hesitate to let me know. I will be look forward to having the chance to serve you.

Best regards,

Linda Chen
Sales Manager

翻譯

我們從 CPU 雜誌得到您的名字，我們也知道您正在尋找矽橡膠墊。我們公司在這個領域已經有超過 20 年的經驗，我們的客戶包括德州儀器、摩托羅拉、日立等。

我們的公司已經獲得德州儀器的頒獎，得到「供應商卓越獎」。另外，Hitach 也認定我們是一個值得信賴的供應商。所有這些證明我們不僅提供最具競爭力的價格，最好的質量和性能，而且還有能力幫助客戶解決他們的問題。

最近，我們擴充了工廠並安裝了更多更新的機器，我們有信心在我們目前的裝備上滿足您的要求。請訪問我們的網站：www.tbi.com.tw，了解更多詳情。如果您對任何產品感興趣，請不要猶豫，讓我知道。我會期待有機會為您服務。

範例 3-6　賣方：透過政府機關協助找到潛在客戶

To: jackturner@piorneer.com
From: lindachen@tbi.com.tw
Cc:
Bcc:
Subject: Silicon Rubber Pads
Attachment: catalog/quotation

Dear Mr. Turner,

Through the recommendation of your Commercial Office here, we got your name and know your interest in our Silicon Rubber Pads, so we are pleased to attach our latest catalog and the most attractive quotation for your reference. For products' more details, you also can visit our company's web at: www.tbi.com.tw. We do believe that our highest quality and competitive price will certainly bring you a considerable and profitable business in your market.

Our company has been engaged in this field for more than 20 years and has earned a very high reputation worldwide. You are assured of our best quality products and the fully support. Don't hesitate to contact us. The faster movement will definitely get you the more market share and profits.

Best regards,

Linda Chen
Sales Manager

翻譯

通過您這裡的商業辦公室的推薦，我們得知您的名字，並知道您對我們的矽橡膠墊的產品有興趣，所以我們很高興附上我們最新的目錄和最有吸引力的報價供您參考。有關產品的更多詳情，您也可以訪問我們公司的網站：www.tbi.com.tw. 我們相信，我們的高品質和有競爭力的價格肯定會為您帶來在您的市場中可觀的和有利可圖的業務。

我們公司在這個領域已有 20 多年的歷史，並在全球贏得了很高的聲譽。 您可以確信我們有最優質的產品和全力支持。請不要猶豫與我們聯繫。快速的行動一定會讓您獲得更多的市場占有率和利潤。

範例 3-7　賣方：雙方朋友介紹找到潛在客戶

To	jackturner@piorneer.com
From	lindachen@tbi.com.tw
Cc	
Bcc	
Subject	LED products
Attachment	catalog

Dear Mr. Turner,

Mr. Mike Jorden, our mutual friend, gave me your name and recommended that you are a powerful importer of the LED products in your country. By taking this chance, I would like to introduce our latest products to you.

Enclosed please find our latest catalog showing you the most popular styles this year. However, if you cannot find any item to satisfy your requirements, please send us your drawing/sketch or picture for the style you prefer, we will offer our best price and sample to you for approval within the shortest time, as we also accept custom-made products.

We assure you our excellent quality, and are looking forward to starting mutually benefit business relationship soon.

Best regards,

Linda Chen
Sales Manager

翻譯

我們共同的朋友 Mike Jorden 先生給了我您的名字，並向我推薦您，說您是貴國 LED 產品的最重要進口商之一。藉此機會，我想向您介紹我們的最新產品。

隨函附上我們最新的產品目錄，向您展示我們今年最流行的款式。但是，如果您找不到任何物品來滿足您的要求，請將您的圖紙／草圖或圖片寄給我們，以獲得您最喜愛的樣式，我們會在最短的時間內為您提供最優惠的價格和樣品，我們可以接受客製化的定製產品。

我們向您保證我們有最優秀的品質，並期待儘快啟動雙方互惠互利的業務關係。

常用句

1. We are happy to advise you that we have lunched new products.
2. You will be interested to know that we have just introduced our new product.
3. We have enclosed our latest catalog.
4. We have pleasure in sending our catalog and price list of newly developed products.
5. We are pleased to introduce our new product and feel sure that you will find it very interesting.
6. We are able to offer you very favorable prices on some goods we have recently been able to purchase.
7. We hope you will take fully advantage of this exceptional offer.
8. We feel sure you will find a ready sale for this excellent material and that your customers will be well satisfied with it.
9. We feel sure you will agree that this product is not only of the highest quality but also very reasonably priced.

練習題

1. 將下列的句子翻成英文

A. 想推薦您一項新家具，該產品目前在美銷售炙手可熱。

B. 我們的產品材質最好，技術最佳，在設計上及信用上無人能出其右。

C. 請注意本項新商品，本產品在今年的國際商展上極為轟動。

D. 您將會很熱衷於內附的一份完整的目錄，為本公司為聖誕節及新年所開發的新玩具。

E. 對於我們剛上市的微波爐，您一定有興趣了解。

F. 附上銷售資料一份，您一定能了解我們的新產品已得到全球的熱烈迴響。

G. 由 Trade Sources 得知貴公司大名並獲知貴公司是貴國主要的電腦產品進口商之一。

H. 作為一家經驗豐富的出口商，本公司自 1980 年以來一直從事電腦產品產業，並和本地廠商關係良好。

I. 本公司為最大的玩具供應商之一，我們有信心令客戶滿意。

J. 經由貴國商會的推薦，我們想知道貴公司是否有興趣進口電腦產品。

2. 將下列中文書信翻譯成英文

我們很高興從漢諾威展覽上得知貴公司是德國主要的電子產品進口商，並希望利用這一次的機會介紹給貴公司我們公司今年最新發展出來的一個高科技的產品。

本公司具有多年的研究發展經驗，並且擁有許多專業的工程師，我們能夠不斷的提供客戶最新的產品，並為客戶創造更好的獲利商機。附件是我們的相關目錄與報價單僅供給您參考。

感謝您的興趣並希望很快可以與您建立雙方互利的商業關係。

3. 請自由發揮寫一封完整的鞋子的推銷信，註明我們的工廠在大陸可以提供價廉物美的商品，消息來源設定為從對方國家的商務部。

4. 寫一封參展【台北文具春季展】後的推銷信，對換過名片的潛在客戶去函告知寄給對方我們的商品型錄。

5. 請寫一封推銷信，信中說明你目前在一家航空公司工作，你是從維他命出口的網頁上得知 **Jolie Boutique**。說明此航空公司已成立超過 **8** 年，你的公司有良好的商譽，並且送貨迅速而值得信賴，從未掉過包裹。所以想提供對方一個優惠的運送價格。因為公司與其他運輸業者有良好的商業關係，因此可以提供更低廉的價格。

CHAPTER

04 詢價

Inquiry

詢價信是指買方或是潛在客戶在收到賣方的推銷信或是經由其他管道得知賣方的產品後，有興趣進而聯絡賣方來詢問報價或是價格的信件。詢價信是買方寄出的信，如果不知賣方應該聯絡何人，可將信寄到該公司的業務銷售或是行銷部門。

買方詢問的項目包含很多，例如索取樣品、產品規格、功能、外觀、付款條件、限制範圍、製造時程、交貨期等，包羅萬象。一般說來，目前很多做外貿的公司都有公司網頁，在網頁上客戶可以直接寫詢價函。如果買方是透過其他管道像是同業推薦、雜誌、政府機關推薦、網路搜尋得知賣方的消息，則可寫一封詢價的 e-mail。

詢價信中，務必說明你想要什麼產品或是服務、需要的數量與運送地點以及賣方後續如何與你聯絡。如果賣方有在網頁上提供詢價表格，則可於表格上直接書寫即可。但是如果賣方是世界大廠或是知名度高的公司，或是特殊產品的供應商，那麼詢問函中最好加上自我介紹，這樣賣方會比較願意就詢問的內容回覆給你，否則詢問信很容易就會被丟到垃圾信件匣內 (junk mail)。

範例 4-1　買方：看展後向賣方詢價

To	lindachen@tbi.com.tw
From	jackturner@pioneer.com
Cc	
Bcc	
Subject	Inquiry for LED Products
Attachment	

Dear Linda,

I had the pleasure of meeting you at the Chicago International Fair last month. I thank you for giving me the information that I asked for.

I am very interested in knowing more about your products, especially the ones we discussed at the fair. Please kindly give us more information, for example: specification, function, limitation and price to L.A. If possible, please send 5 samples to me, our company can analyze everything in detail.

I hope that we can cooperate with each other in the near future.

Best wishes,

Jack Turner
Purchasing Manager

翻譯

我很高興在上個月的芝加哥國際商展上與您見面。感謝您回覆了我向您詢問的事情。

我非常有興趣了解更多關於您的產品的信息，特別是我們在展會上討論的產品。請提供更多信息，例如：規格、功能、限制和到美國洛杉磯的價格。如果可能的話，請寄5件樣品給我，我們公司可以詳細分析一下樣品。

希望在不久的將來我們能夠相互合作。

範例 4-2　買方：參觀商展後向賣方詢價

To lindachen@tbi.com.tw

From jackturner@pioneer.com

Cc

Bcc

Subject Inquiry for LED Products

Attachment

Dear Linda,

Following our conversation with the representatives in your stand at the Chicago International Fair, I would be grateful if you can send me your latest catalog of LED products.

As there is a large demand for this new product in our market, we are willing to import this item recently. In the meantime, we must point out that as the possible buyers concern prices, your quotation must be competitive.

If you can guarantee prompt delivery and can offer the best price, I may be able to place an order for 3,000 sets. My credit references will be provided when I make an initial order.

Best wishes,

Jack Turner
Purchasing Manager

翻譯

在我與你們在芝加哥國際商展的業務代表交談之後，如果你能寄給我你們最新的 LED 產品目錄，我將不勝感激。

由於我們市場對這種新產品的需求很大，我們最近願意進口這個產品。同時，我們必須指出，由於我們的客戶會特別注重價格，你的報價必須具有競爭力。

如果您能保證可以及時交貨並能提供最優惠的價格，我可以下訂單 3,000 組。我們在下第一張訂單時，將提供信用諮詢人以供備詢。

範例 4-3　買方：經政府機構得知賣方而來詢價

To sales.dept@tbi.com.tw

From jackturner@pioneer.com

Cc

Bcc

Subject Inquiry for Auto parts

Attachment

Dear Sir,

Your company was given by Taiwan Trade Center in Panama as a manufacturer and exporter of auto parts. We take this opportunity of writing to you to introduce ourselves.

Our company is actively operating as an importer and distributor of auto parts. We are sourcing new suppliers for good quality and low prices auto parts and accessories. Please kindly contact us and send us your catalog and price list.

Thank you in advance and I look forward to your reply soon.

Best wishes,

Jack Turner
Purchasing Manager

翻譯

我們是經由巴拿馬的台灣貿易中心得知貴公司，知道貴公司是汽車零件的製造商和出口商。我們藉此機會寫信給您自我介紹一下。

我們公司是汽車零件的進口商和經銷商。我們正在尋找可以提供優良品質和競爭價格的汽車零件供應商。請與我們聯繫，並寄給我們最新目錄和價目表。

在此先感謝您，期待您的回覆。

範例 4-4　買方：經雜誌得知賣方而來詢價

To	sales.dept@tbi.com.tw
From	jackturner@pioneer.com
Cc	
Bcc	
Subject	Inquiry for Fans
Attachment	

Dear Sir,

We have seen your advertisement in "The Houseware" and are interested in your ceiling fans and stand fans.

Please kindly quote your best prices based on FOB Taiwan, stating your earliest delivery date, your terms of payment and discounts for regular purchases.

We look forward to creating a partnership that is beneficial to us both.

Best wishes,

Jack Turner
Purchasing Manager

翻譯

我們在《家庭用品》的雜誌上中看到您的廣告，並對您的吊扇和立扇感興趣。

請根據「台灣離岸價」報給我們最好的價格，並說明您最早可以交貨的日期還有付款條件，並告知常態性的訂單的折扣是如何計價。

我們期待建立一個雙方都有益的合作夥伴關係。

範例 4-5　買方：經雜誌得知賣方而來詢價，詳列詢價項目與買方聯絡方法

To	sales.dept@tbi.com.tw
From	jackturner@pioneer.com
Cc	
Bcc	
Subject	Inquiry for Kitchen Knife
Attachment	

Dear Sir,

We are one of the leading companies here in Germany in dealing with kitchen supplies. We have seen your advertisement in Appetite and are interested in your variety of utensils, particularly knives.

We would appreciate a quote for the entire Blumenschein Large Knife collection, the Braun Standard Kitchen line and the Modern Quality Chef's set. Please indicate prices C.I.F. Kaohsiung, Taiwan. Please indicate your earliest delivery date, terms of payment, and discounts for regular purchases. We would also appreciate receiving your catalog.

Best wishes,

Jack Turner
Purchasing Manager

翻譯

我們在德國是廚房用品的領導公司之一。我們在 Appetite 上看到了您的廣告，並對您的各種餐具特別感興趣，特別是刀具。

我們希望您可以報價給我們有關整個 Blumenschein 大刀具系列，Braun 標準廚房系列和現代專業廚師系列產品。請註明報價是 C.I.F. 高雄，台灣，並請說明您可以的最早交貨日期，付款條款和常態訂單的的折扣條件。我們將很感謝您寄給我們目錄。

範例 4-6　買方：就產品詢問報價

To sales.dept@tbi.com.tw

From jackturner@pioneer.com

Cc

Bcc

Subject Request for quotation

Attachment

Dear Sir or Madam:

Thank you for your attention. I am the purchasing manager for Pioneer Corporation, US. I would like to request a quotation for the following products.

- One 24-track hard disk recorder
- One 8-channel MIDI fader
- One 8-channel control surface extender

Please feel free to contact me with any questions at 1-408-576-8888 ext. 234 or over email at jackturner@pioneer.com.

Best wishes,

Jack Turner
Purchasing Manager

翻譯

您好：我是美國先鋒公司的採購經理。想就下列產品詢問報價。

- 一個 24 音軌硬碟錄音機
- 一個 8 頻道樂器數位介面推桿
- 一個 8 頻道控製面板擴充器

如有任何疑問，請隨時與我聯繫 1-408-576-8888 分機 234 或通過電子郵件 jackturner@pioneer.com。

常用句

1. I would like to request your latest catalog.
2. I am interested in receiving more information about your new product.
3. We are interested in your products. Do you have any brochures which you could send us?
4. We are pleased to get in touch with you regarding the supply of stainless steel.
5. Would you please send me your updated catalog?
6. Please send a catalog of your newest products.
7. We are looking for the hand tools and please advise if you are able to supply this item.
8. Please send us samples of the goods.
9. I was interested to see your advertisement for ××× product.
10. I understand you are manufacturers of ××× and should like to receive your current catalog.
11. Please advise whether you can supply the goods from stock as we need them urgently.
12. Kindly send us without charge some samples of your line.
13. If you can supply suitable goods, we may place regular orders for large quantities.
14. If you are interested in doing business with us, please inform us of your terms and conditions of business.

練習題

1. 將下列的句子翻成英文

A. 請寄上最新的產品目錄。

B. 請寄上一些免費的樣品。

C. 如蒙寄上貴公司最新開發的新品目錄，將感激不盡。

D. 我們聽聞貴公司已推出一種新的電動車，我們想知道進一步的細節。

E. 我們對您的產品有興趣，請寄上產品目錄。

F. 請提供目錄及報價單以供評估。

2. 將下列中文書信翻譯成英文

我們從遠東貿易辦事處得知貴公司的大名，並有興趣貴公司的產品。

我們是美國的主要運動鞋進口商之一，本公司有超過 500 個業務，也有很好的銷售管道，和買主都有很好的關係。因此，請寄給我們最好的報價單，並註明不同訂購量的折扣狀況。

謝謝您的協助，靜候回音。

3. 請寫封詢價信函（感謝賣方寄來的目錄及公司光碟，研讀後對 MRI 系列極有興趣，請賣方報最好價格，並請寄一套樣品來）。

4. 寫一封詢問函，請求賣方提供最新的商品型錄與報價。（產品自定）

5. 寫一封詢價信函，表示在雜誌「The Houseware」上看到賣方的商品廣告，我們對日式茶壺和碟子有興趣。請賣方報價 CIF LA 及價格折扣、出貨日與付款方式，並請提供一套樣品，以供評估。

CHAPTER

05 報價

Quotation

5.1 被動式的報價

報價可分為被動式的報價與主動式的報價。被動式的報價信 (letter of quotation) 就是對客戶的來函詢價 (inquiry letter) 所作的回覆。通常收到客戶的詢價信後，最好當天或是隔天就應該馬上回覆，以示效率與誠意。因為一般客戶都會同時詢問數家的供應商，如果對客戶的問題中有無法立即回覆的事項，應於回函時告知對方大約可回覆的預計時間，以免客戶等待過久。

目前很多做外貿的公司都有公司網頁，在網頁上客戶可以直接寫詢價信函，賣方可於網頁的詢價信上得知潛在客戶的需求，並立即回覆。或是買方透過其他管道寫一封詢價的信函來，賣方也可針對此來回覆。通常如果買方的詢價項目不多，賣方可直接於 e-mail 上回覆。如果詢價項目很多，有可能是一系列或是相關產品，則賣方應該用附件的方式將所需的產品以報價單的方式報價。報價單的製作是比較正式的方式，通常報價單上會有公司表頭，並且會載明商品型號、單價、交易條件、付款方式、製造所需最短時間 (lead time)、最低訂購量、產品包裝方式、報價有效期間…等。

報價的方式在不同的情況下有不同的報價方式，以下為常見的報價方式：

1. **Selling offer（賣方報價）**：賣方將所有條件（如商品型號、名稱、數量、品質、價格、交貨日期、付款方式、有效期間…等）通知買方，表示願意以此條件將商品售予對方之意。

   (1) Firm offer（穩固報價）：表示承諾以一固定價格，於一定的期間內出售特定商品。對方一旦接受此報價，契約即成立。

   (2) Non-Firm offer（非穩固報價）：表示報價可依情況進行調整，意思就是價格可能還有變動的狀況。因此於報價上會出現下列說明的文字。

   - Offer without engagement（不受約束報價）
   - Offer subject to confirmation（經確認才為有效的報價）
   - Offer subject to goods being unsold（價格以貨物尚未出售才有效）
   - Offer subject to prior sale（價格以貨物尚未出售才有效）
   - Offer subject to market fluctuation（價格依市場狀況而變動）
   - This quotation is subject to our final consideration（價格依我們最後的考量而定）
   - Offer on approval（價格經買方同意而定）

2. **Buying offer（買方報價）**：指買方告知賣方將其希望的購買條件（商品名稱、數量、價格、交貨、攬貨公司、付款條件…等）通知賣方，表示願意以此條件向賣方購買商品。

範例 5-1　賣方：依客戶來函詢問回答（Re 範例 4-1）

To	jackturner@pioneer.com
From	lindachen@tbi.com.tw
Cc	
Bcc	
Subject	Quotation for LED products
Attachment	

Dear Jack,

Thank you for your e-mail dated July 13 inquiring about our LED products. Please find our best quotation for the item in which you expressed interest. We hope that you agree that our offer is very competitive. We also send you 5 samples today upon your request.

Price: US$1.12/pc FOB Taiwan by sea freight
Minimum order quantity: 6000 pcs
Packing: Bulk packing
Shipment: in two weeks after receipt of payment
Payment: by irrevocable L/C at sight in our favor
Validity: 30 days

In addition, we send you today our complete catalogue and brochures by Federal Express. As our items are numerous, please kindly study our catalogue and then inform me of the items you are interested, so that I can provide you our best prices as soon as possible.

Thank you for your attention to the above, and look forward to your business in the near future.

Best regards,

Linda Chen
Sales Manager

整個系列產品報價的提供通常不會一次就給新的客戶，因為怕客戶將此資料洩漏給競爭對手。或是因為項目太多，客戶不見得會全部購買，不必做無效報價，故可以請客戶選擇有可能購買的項目再行報價。

翻譯

感謝您於 7 月 13 日來函詢問我們的 LED 產品。對您表達感興趣的項目我們報給您我們最好的報價。我們希望您同意我們的報價非常具有競爭力。

價格：US $ 1.12 / 件 FOB 台灣離岸價／海運
最小訂購量：6000 個
包裝：散裝包裝
裝運：收到付款後的兩週內
付款方式：以我們為受益人的不可撤銷信用狀
有效期：30 天

此外，我們今天也以 FeDex 寄給您我們完整的目錄和手冊。由於我們的項目很多，請您先看看我們的產品目錄，然後告訴我您感興趣的產品，以便我儘快為您提供最優惠的價格。

感謝您，期待未來能收到您的訂單。

範例 5-2　賣方：依客戶來函詢問回答（Re 範例 4-2）

To	jackturner@pioneer.com
From	lindachen@tbi.com.tw
Cc	
Bcc	
Subject	Quotation for LED Products
Attachment	catalogue

Dear Jack,

Thank you for your email dated Aug. 8 inquiring about our LED products. As requested, we enclose herewith an image of our standard LED bulb and quote as below:

Part No.: N-13S LED bulb
FOB price: US$0.15/pc
Min. Q'ty: 5,000 pcs
Shipment: within 30 days after order confirmed
Payment: by T/T before shipment
Validity: 30 days from the date quoted

Please check if the above meet your requirement, if not, please send us your image or picture by email for quoting.

Thanks for your kind attention and look forward to hearing from you soon.

Best regards,

Linda Chen
Sales Manager

翻譯

感謝您 8 月 8 日的來函詢問我們的 LED 產品。根據要求，我們隨函附上我們標準 LED 燈泡圖片和報價如下：

項目編號：N-13S LED 燈泡
FOB 價格：每個 $0.15 美元
最低數量：5,000 個
交期：訂單確認後 30 天內
付款方式：出貨前以電匯付款
有效期：自報價日算起 30 天內有效

請檢查以上是否符合您的需求，如不符合，請以電子郵件傳來貴公司的圖片或是照片以利報價。

感謝您的關注，並期待儘快收到您的回音。

範例 5-3　賣方：依客戶來函詢問回答（Re 範例 4-3）

To jackturner@pioneer.com

From lindachen@tbi.com.tw

Cc

Bcc

Subject Quotation for auto parts

Attachment catalogue

Dear Mr. Turner,

Thank you for your inquiry for our products, herewith we will quote you our best sellers as shown below:

	FOB Taiwan
Item No. 3751 headlight	US$8.36/pc
Item No. 4513 seat	US$3.52/pc
Item No. 6271 mirror	US$0.53/pc

Packing: 1pc/poly bag, 50pc/box, 200pc/carton
(NW: 14kgs; GW:15kgs; Measurement: 0.8cuft for each carton)
Minimum: 5,000 pcs per item; US$6,000 per shipment

These auto parts have been selling extremely well and we recommend them to you with confidence. There are many manufacturers who are making these products. However, our quality is superior to other suppliers'. We would recommend you to place your order as soon as possible as we have been receiving many rush orders recently.

As requested, I also attached our latest catalog for your reference. If you are interested in other items besides above quoted items, please let me know. I will give you a quotation immediately.

Any order that you might give us will receive our best attention and prompt service. We look forward to receiving your trial order soon.

Best regards,

Linda Chen
Sales Manager

翻譯

感謝您對我們產品的諮詢，以下的報價是我們目前最暢銷的產品：

	FOB 台灣
產品編號 3751 頭燈	US $ 8.36 ／件
產品編號 4513 座位	US $ 3.52 ／件
產品編號 6271 鏡子	US $ 0.53 ／件

包裝：1 件／袋，50 件／盒，200 件／箱
（每箱淨重 14 公斤；毛種 15 公斤；才數 0.8 立方呎）
最低數量：每項產品 5,000 件 ; 每批貨物 6,000 美元

這些汽車零件銷售非常好，我們非常推薦這些產品。目前雖然有許多製造商正在製造這些產品。但是，我們的品質是優於其他供應商。我們建議您儘快下訂單，因為我們近期收到很多緊急訂單。

根據您的要求，我們附上了我們最新的產品目錄供您參考。如果您對除上述商品之外的其他商品感興趣，請告訴我。我會立即報價給您。

我們對於您的任何訂單都會特別關注和立即的服務。我們期待儘快收到您的訂單。

範例 5-4　賣方：依客戶來函詢問回答（Re 範例 4-4）

To jackturner@pioneer.com
From lindachen@tbi.com.tw
Cc
Bcc
Subject Quotation for fans
Attachment

Dear Jack,

Thank you for your email dated on Feb. 22 expressing your interest in our ceiling and stand fans. Regarding your inquiry, following are our answers:

	FOB Taiwan
Ceiling fan	US$12.75/pc
Stand fan	US$11.63/pc

Delivery: within 45 days after receipt of your order
Payment: by irrevocable L/C at sight in our favor or T/T before shipment
Minimum quantity: 5,000 pcs per shipment or US$4,500 per shipment. Otherwise, we must request US$250 as handling charge to cover our cost for inland charges, taxes and other expenses

For regular order, if your order quantity reaches 500,000 pcs, we will offer you 2% discount; 1,000,000 pcs, will reach 3% discount. Quantity over 1,000,000 pcs will be granted 5% which is our maximum discount rate.

We look forward to your comments or your first order confirmation soon.

Best regards,

Linda Chen
Sales Manager

翻譯

感謝您 2 月 22 日的電郵來函表達您對我們的吊扇和立扇的興趣。 關於您的詢價，以下是我們的報價：

	FOB 台灣
吊扇	US $ 12.75 / 件
立扇	US $ 11.63 / 件

交貨：收到訂單後 45 天內
付款方式：以我們為受益人的不可撤消的信用狀或是出貨前電匯
最小數量：每次出貨 5,000 件或每次出貨 4,500 美元。
否則，我們將會收取 250 美元的手續費來支付內陸費用，稅費和其他費用

對於常態性的訂單，如果您的訂單數量達到 500,000 個，我們提供給您 2%的折扣；1,000,000 個，將有 3%的折扣。 超過 1,000,000 件的數量將有 5%，這是我們的最大折扣率。

我們期待儘快收到您的意見或是第一份訂單確認。

範例 5-5　賣方：依客戶來函詢問回答（Re 範例 4-5）

To jackturner@pioneer.com

From lindachen@tbi.com.tw

Cc

Bcc

Subject Quotation for kitchen knives

Attachment Q52133.doc

Dear Mr. Turner,

Thank you for your email dated Apr. 16, introducing yourself and inquiring about the items in which you are interested.

Attached please find our best quotation Q52133 for your reference. We hope our offer can meet your demand so that you may confirm your order in the near future. We also send you our latest catalog by UPS today.

We are glad to have this opportunity to serve you and hope that we can cooperate together very soon.

Best regards,

Linda Chen
Sales Manager

翻譯

感謝您在 4 月 16 日的來函介紹自己並詢問您感興趣的商品。

附件是我們的報價單 Q52133 供您參考。我們希望我們的報價能夠滿足您的需求，以便您可以在不久的將來下訂單。另外，今天我們用 UPS 寄給您我們最新的目錄。

我們很高興能有這個機會為您服務，希望未來我們能夠很快合作。

附件：報價單

Taiwan Bear International

3F, No. 11, Park Avenue II
Science-Based Industrial Park
Hsin-Chu 30075, Taiwan
Tel: 886-03-5798888
Fax: 886-035978891

QUOTATION

ATTN：Mr. Turner

No: Q52133
Date: Apr. 15, 2021

P/N	Description	Min. Q'ty	Unit Price
			CIF Taiwan
	Blumenschein Large Knife Line		
5110	Bread knife	1,000 pcs	US$13.80/PC
5120	Chef's knife	1,000 pcs	US$8.50/PC
5130	Paring knife	1,000 pcs	US$5.25/PC
	Braun Standard Kitchen Line		
2185	Bread knife	3,000 pcs	US$8.35/PC
2260	Chef's knife	3,000 pcs	US$6.20/PC
2796	Paring knife	3,000 pcs	US$4.10/PC
6897	Modern Quality Chef's Set	3,000 pcs	US$25,50/PC

Packing: standard export packing
Shipment: within 45 days after receipt of order
Payment: by irrevocable L/C at sight in our favor
Validity: 30 days

範例 5-6　賣方：就客戶來函詢問的產品報價（Re 範例 4-6）

To jackturner@pioneer.com
From lindachen@tbi.com.tw
Cc
Bcc
Subject Quotation
Attachment

Dear Mr Turner,

Many thanks for your inquiry of our products and are glad to quote you our best price and terms as below:

QUOTATION

Item	Description	Min. Q'ty	Unit Price (FOB TWN)
121H	One 24-track hard disk recorder	1000	US$6.35
556B	One 8-channel MIDI fader	1000	US$5.85
874A	One 8-channel control surface extender	1000	US$7.52

Packing: standard export packing
Delivery: about 30 days after receipt of order
Payment: by irrevocable L/C at sight in our favor
Validity: 30 days or subject to our final confirmation when ordering

We hope you are satisfied with our offer and can confirm the terms you prefer. In expectation of your trial order confirmation in return, we thank your for your attention and remain.

Best regards,

Linda Chen
Sales Manager

翻譯

非常感謝您對我們產品的詢問，我們非常樂意提供我們最好的報價和條件給您：

報價單

項目	描述	最小數量	單價 (FOB TWN)
121H	一個 24 音軌硬碟錄音機	1000	US$6.35
556B	I 一個 8 頻道樂器數位介面推桿	1000	US$5.85
874A	I 一個 8 頻道控製面板擴充器	1000	US$7.52

包裝：標准出口包裝
交期：收到訂單後約 30 天
付款方式：以我們為受益人的不可撤銷信用狀
有效期：30 天或在訂購時由我方做最終確認

我們希望您對我們的報價感到滿意，並確認您可接受的條款。
在此感謝您對我們商品的關注並期待能收到您的訂單。

5.2 主動式的報價

主動式的報價通常是對目前有業務往來的客戶或是目標 (target) 客戶主動去函報價。一般對目前有銷售給客戶的商品，最少會一年更新一次價格。價格更新一定要提前通知客戶，一般最好是一季以前通知，這樣客戶比較有充裕的時間應對。若是有曾經往來的客戶，但是有一段時間卻沒有再下新的訂單，我們也可以藉由新產品上市再主動去追蹤與報價，以了解客戶的狀況。

另外，有時候我們會不定期更新某些商品的價格。這種情況有可能是受外在環境的影響，像是石油上漲造成的物價上漲、工資或是原物料上漲、物價波動太大，或是因匯率波動很大而影響價格，這時候我們就會調整商品價格並主動報價給客戶。

範例 5-7　賣方：主動報價給舊客戶

To	jackturner@pioneer.com
From	lindachen@tbi.com.tw
Cc	
Bcc	
Subject	Quotation
Attachment	pricelist.doc

Dear Jack,

Please refer to our new launched product P/N:7381AR. We are pleased to inform you about the great interest aroused in this product at the recent Chicago International Trade Fair. Numerous inquiries followed this exhibition. Since you are one of our best and important customers, we would like to introduce to you this new product and give you our best offer.

This new product is slim and compact, weighing no more than 200 grams and measuring 10 cm x 8 cm x 2 cm. With a memory of 128G, you can store as many as 10 films at one time. Also, it can be connected to the internet, retrieve e-mails, and functions more or less like a cell phone. It comes with a standard high-resolution LCD monitor, displaying true colors. The market demand is increasing now.

With this e-mail, we enclosed details of our export prices and terms. We are pleased to inform you that we can supply any quantity straight from stock. Delivery is guaranteed with one week. Please take action, send your order as early as possible. This product will be very favorable for your market.

We look forward to receiving your order soon.

Best regards,

Linda Chen
Sales Manager

通常有新產品上市時，需即時通知舊客戶。一般舊客戶的價格需於一定期間更新，最長時間不會超過一年。一般碰到物價波動很大或是匯率波動很大時，賣方會不定期更新售價。

翻譯

請參考我們新推出的產品，編號：7381AR。我們很高興地通知您最近在芝加哥國際貿易商展上這項產品引起了很多採購的極大興趣。在本次展覽我們接到大量的詢問函。由於您是我們最重要的客戶之一，我們想向您介紹這款新產品，並提供給您我們最好的報價。

這款新產品纖細小巧，重量不超過 200 克，尺寸為 10 公分 ×8 公分 ×2 公分。內建 128G 的記憶體，您可以一次存儲多達 10 部電影。此外，它可以連接到聯路上，下載電子郵件，功能與手機相同。它配備了一個標準的高解析液晶顯示器，顯示真實的色彩。目前市場對此商品的需求正在增加。

我們隨函附上了我們的報價和條款細節。我們很高興地通知您，我們目前可以直接提供您所要求的任何數量。交期保證可以在一週內出貨。請立刻採取行動，儘早下訂單給我們。這個產品將會非常有利於您的市場銷售。

我們期待儘快收到您的訂單。

範例 5-8　賣方：通知舊客戶因原物料成本增加而漲價

To	jackturner@pioneer.com
From	lindachen@tbi.com.tw
Cc	
Bcc	
Subject	Price Increase
Attachment	pricelist.doc

Dear Jack,

Please be informed that our item no. MN22135-B will be adjusted in price from July 1st, 2021 as the cost of raw materials has risen continuously.

Old price:　US$50.25/set
New price:　US$52.35/set

Therefore, please kindly consider placing your order before July, the price will remain the old one. We will appreciate if you can confirm the quantity in 2 weeks, so that we can book materials at once in order to avoid another increase.

We look forward to your order confirmation soon.

Best regards,

Linda Chen
Sales Manager

翻譯

我們在此通知您我們的產品編號 MN22135-B，因為原材料成本持續上漲，將於 2021 年 7 月 1 日起調整價格。

原價格：50.25 美元／套
新價格：52.35 美元／套

因此，敬請考慮在 7 月之前下訂，因為 7 月前下的訂單，價格將保持舊價格。如果您能在兩週內確認數量，我們將非常感激，因為這樣我們可以一次預訂材料以避免成本再次增加。

我們期待您的訂單確認。

範例 5-9　賣方：通知舊客戶商品全面漲價

To	jackturner@pioneer.com
From	lindachen@tbi.com.tw
Cc	
Bcc	
Subject	Price Increase
Attachment	

Dear Jack,

Please be advised that effective from May 1st, 2021, all the prices we offered you will increase 10% for all items. Due to increase of wages and raw materials for the past few months, we are compelled to adjust our prices to cover at least part of this additional costs. Please kindly understand.

The updated price list will be sent before March 15. The old prices will be maintained until end of April. Please kindly consider placing orders before April to avoid a price increase next quarter.

Best regards,

Linda Chen
Sales Manager

請注意，從 2021 年 5 月 1 日起，我們提供的所有商品價格將調漲 10%。由於過去幾個月工資和原材料成本增加，我們不得不調整我們的價格以至少支付這部分額外的漲價。敬請諒解。

更新後的價目表將在 3 月 15 日前寄出給您。舊價格有效期將維持至 4 月底。敬請考慮在四月底前下訂單，以避免下一季的價格調漲。

範例 5-10　賣方：通知舊客戶商品全面漲價（含匯率因素）

To: jackturner@pioneer.com

From: lindachen@tbi.com.tw

Cc:

Bcc:

Subject: Price Increase

Attachment:

Dear Jack,

This is to inform you that from the 1st day of next month, the price of all our products will be raised by 5%. We feel sorry for this increase, but this price increase has become unavoidable in view of the rising costs of labor and raw materials. In addition, our NT dollars is appreciating continuously. New price lists are being prepared and these will be sent to you when they are ready in one week. We will remain our old prices on all orders received here up to and including March 31. Please kindly note and understand.

We appreciate your past favor and look forward to serving you as usual.

Best regards,

Linda Chen
Sales Manager

翻譯

本函是通知你，從下個月一號開始，我們所有產品的價格將上漲 5%。我們對這一漲價感到抱歉，但鑑於人工成本和原材料成本上漲，價格調漲已是不可避免。另外，新台幣的不斷升值也是造成漲價的因素。新的價目表正在準備中，這些價目表將在一周內準備好後寄送給您。在 3 月 31 日以前收到的所有訂單，我們將保持原價。請注意並體諒。

我們感謝您過去的關照，並期待像往常一樣為您服務。

常用句

1. As requested, we quote our best price at FOB Taiwan as the quotation attached.
2. It is our pleasure to learn that your company is interested in dealing business with us.
3. We have quoted our best terms in the enclosed price list and trust that you will agree that our prices are competitive.
4. It is the best offer we can make at present and we trust that it should be acceptable to you.
5. Our prices are subject to change without notice.
6. All prices are subject to market fluctuations.
7. We strongly advise you to take advantage of this exceptional opportunity.
8. If you miss this opportunity, you may not be able to obtain the item even at a higher price.
9. It is certain that these things will advance in price before long, so that we heartily advise you to buy as much as you can.
10. As you will know, the prices have risen sharply during the last few month. So this is a very moderate increase, and we hope that you will take advantage of our offer.
11. As orders have rushing in recently, please place your order as soon as possible.
12. In view of the heavy demand for this article, we would advise you to place your order as early as possible.
13. We look forward to your further inquiries or comments.
14. We look forward to your trial order soon.
15. We look forward to the opportunity of being of service to you.
16. We look forward to your answer and a pleasant business relationship.
17. Your prompt order confirmation will be highly appreciated.
18. You are sure of being satisfied with our products and service.
19. We would recommend you to place orders immediately.

練習題

1. 將下列的句子翻成英文

A. 我們以非常優越的付款條件，D/A 30 天供貨。

B. 我們以美元 30~33 之間的價格供貨。

C. 數量超過 1000 件，可按價格表的價格打 95 折。

D. 由於日用品價格大幅下跌，我們將以低於市價的 15% 價格提供商品。

E. 以下為我們的商品報價，此價格以在 4 月底前收到您的回函為有效。

F. 如您錯失此次良機，以後可能以高的價格都無法買到。

G. 如您所了解在過去幾個星期內物價飆漲得很厲害，因此本價格只是溫和上漲，希望您利用我們給您的價格。

H. 建議切勿錯失良機，保握這次報價。

I. 感謝您 9/28 的詢價，我們很高興報價如下：

J. 以上報價是淨價，不含我方的佣金。

2. 將下列中文書信翻譯成英文

感謝您 3/2 日的來函，詢問我們的毛衣產品。根據您的要求，我們的報價如下：

價錢：每打美金 35.60 元 FOB 台灣離岸價格
交期：收到訂單後 30 天出貨
付款：不可撤銷保兌即期信用狀
最低訂購量：1000 打
有效期：30 天

如需進一步資料，請不吝來函告知。我們靜候貴公司的訂單。

3. 在信件中告訴客戶，報價單如附件，請客戶參考，並告知意見或下訂單。
4. 寫一封在 e-mail 中直接報價給新客戶的信函，再次介紹公司，強調產品的優點，希望客戶能有興趣並儘早下單。
5. 感謝客戶參展時的訪問，附上當時所談商品的報價單，希望客戶及早訂購。
6. 感謝客戶來函表示對我方產品的興趣即索取報價。但是由於我方商品項目繁多，請客戶先於公司型錄上挑選有興趣的產品，我方再進一步報價。
7. 漲價通知，發函給舊客戶。由於工資、物料的上漲，以及台幣的升值，我方價格將於 9 月 1 日起全面調漲 3%。
8. 告訴客戶無法報價，因為 AA 系列已經停產，目前的新商品是 BB 系列，我們可以報 BB 系列的價格，希望客戶可以接受。

CHAPTER

06 還價

Counter Offer/ Negotiation

在賣方報價提出交易條件後，通常買方可能無法全面的滿意賣方的條件。此時，買方對於賣方報價上的商品品質、價格、數量、出貨、付款條件等提出自己的看法與條件與賣方進行協商，此一過程即稱為還價 (counter offer/negotiation)。賣方就買方提出的修改條件，提出賣方可接受的條件後，再一次回覆買方，此一程序也稱為還價。在雙方都同意的條件下，則買賣的商業條件即成立，買方可開始進行購買。

買方既回來還價，就表示買方的購買意願相當高，只是想藉由殺價讓自已的利潤增加。因此，就算賣方的利潤已經不高，賣方應在協商的過程中最好意思意思一下，但也可以將還價的條件再刺激買方儘速下單，例如：第一次的下單需於一星期內可獲得的折扣，或是提高訂購量可給予多一點折扣等方案。切記不可因買方的還價，又急於做成生意就降價太多，這樣會讓買方誤認賣方的利潤很高，就會常常來要求降價。

在國際的商業行為上，常常需經過多次的溝通與交涉，更會有多次的往返交涉，反覆的還價，才能達最後的共識，簽下協定。因此，還價是國際貿易上非常重要的一環，其中涉及了許多策略與談判的技巧，雙方皆會想盡辦法，利用各種方法來維護自已的最大利潤。

賣方在收到買方的還價之後，在回覆買方的過程中，可分為下列幾種情況：

1. 接受買方所有還價條件

對買方的還價，即使是賣方可以接受的範圍，也不要答應得太快。通常賣方應宣稱自己已無利潤空間，或是跟買方說他的還價已低於我們的成本，但是考量對方是新客戶，可能不知我方產品的品質才會殺價殺這麼低，為了讓買方進一步了解我們，我們才願意接受買方得還價來建立雙方的貿易關係。如果是舊客戶或是大客戶的還價，賣方接受其條件，可以宣稱是對老客戶與大客戶的特別優待。

2. 不接受買方還價

如不接受買方的還價，應告知其原因，請買方再次考量一下。原因可能有：我方的品質比其他廠商好 (good quality)、材料成本高 (costly material)、原物料價格上漲 (raw material cost has been raised)、人工成本上漲 (labor cost has been increased)、新台幣升值 (NTD has been appreciated)、匯率浮動太厲害 (fluctuation of the exchange rate)、買方還價低於我方的成本…等理由，因而無法接受。

3. 條件式接受

表示願意接受買方的還價，但是有附帶條件，例如請買方增加購買的數量、改變付款的條件（將信用狀付款改成用電匯事先付款，或是先預付訂金等）、或是將多批出貨改成貨櫃出貨，或是建議買方買較便宜的替代品。

4. 折衷還價

賣方有時為了表示與買方建立商業關係的誠意，雖然買方的還價太過，賣方無法接受買方要求的條件，但是可以折衷的降一價格，例如買方可能要求賣方再降 10% 的折扣，賣方可以折衷還價，表示只願意接受 3% 的還價，再次進行協商。

範例 6-1　買方：客戶抱怨報價太高

To	lindachen@tbi.com.tw
From	jackturner@pioneer.com
Cc	
Bcc	
Subject	Price for beach sandal
Attachment	

Dear Linda,

Thanks for your quotation dated Jan. 25 for the beach sandal. We find your price is 10% higher than other suppliers. Could you re-quote us a better price as we are interested to buy 40,000 pairs a year?

Best wishes,

Jack Turner
Purchasing Manager

感謝您 1 月 25 日的沙灘涼鞋報價。但是我們發現您的價格比其他供應商高出 10%。您可以再重新報一個更好的價格嗎？因為我們有興趣每年購買 40,000 雙。

範例 6-2　買方：客戶還價

To	lindachen@tbi.com.tw
From	jackturner@pioneer.com
Cc	
Bcc	
Subject	Your offer of Q1220
Attachment	

Dear Linda,

We received your quotation of Q1220, but we found that you raised the price a lot. However, we still tried very hard to negotiate with our customer with your new offer.

According to our previous deals, you always granted us 10% discount from your listing price. Therefore, we offered the same discount to our customers and have already got orders from customers.

Please kindly confirm your acceptance to keep the previous terms. Thanks for your help in advance.

Best wishes,

Jack Turner
Purchasing Manager

翻譯

我們收到了您的報價單 Q1220，但我們發現價格漲了很多。雖然如此，我們仍然非常努力地與我們的客戶協商您的新報價。

根據我們之前的協議，您給我們的報價是您售價的九折。因此，我們也是以此基準向客戶提供報價，並已獲得客戶的訂單。

請確認這樣的條件沒有改變。感謝您的協助。

範例 6-3　買方：客戶告知購買量要求重報價格

To	lindachen@tbi.com.tw
From	jackturner@pioneer.com
Cc	
Bcc	
Subject	Item no. A5789
Attachment	

Dear Linda,

Thank you very much for your prompt reply. Our customer has come back to us saying that they cannot take 10,000 sets of kitchen knife (item no. A5789). Please re-quote price based on 2,000 sets.

Also, please be noted that your previous quotation is not competitive enough as we almost could not get any profits on this base. We expect you can offer a better price this time. Thanks for your support.

Best wishes,

Jack Turner
Purchasing Manager

翻譯

非常感謝您的及時回覆。我們的客戶回來告訴我們，他們無法一次購買 1 萬套菜刀（貨號 A5789）。請重新以出貨 2,000 套報價。

另外，請注意，您之前的報價沒有足夠的競爭力，因為我們幾乎無法在此報價上獲得任何利潤。我們希望這次您可以提供更好的價格。感謝您的支持。

範例 6-4　買方：客戶告知報價太高，需重報並大幅降價才有機會接到訂單

To	lindachen@tbi.com.tw
From	jackturner@pioneer.com
Cc	
Bcc	
Subject	Item no. A5789
Attachment	

Dear Linda,

We have noticed you that your quotation is much higher than that of the other suppliers. We have found out the price difference is about US$1.2/pc.

Our customer would like to give us the opportunity to get this business because of our long-term relationship. Therefore, your price must be “within reason” or somewhat in line with the competition.

The order quantity will be 15,000 which will cost customer to pay extra US$18,000 (US$1.2×15,000). With such extra cost, we will definitely not get this order. We suggest you to review your quotation carefully and give us an explanation or re-quote a competitive price.

Best wishes,

Jack Turner
Purchasing Manager

翻譯

我們已經注意到您的報價遠高於其他供應商的報價。我們發現價格差異約為 1.2 美元／件。

由於我們與客戶的長期合作關係，我們的客戶希望給我們這個機會獲得這項業務。因此，您的價格必須「合理」或是能與與競爭對手相比。

訂單數量是 15,000 件，這將讓客戶額外支付 18,000 美元（1.2 美元 ×15,000 美元）。有了這樣的額外費用，我們絕對不會得到這個訂單。我們建議您仔細查看您的報價，並給我們一個解釋或重新報一個有價競爭力的價格。

範例 6-5　買方：客戶告知因市場競爭激烈，請不要漲價太多

To lindachen@tbi.com.tw

From jackturner@pioneer.com

Cc

Bcc

Subject Your offer dated on 5/21

Attachment

Dear Linda,

Please note that it is very difficult to promote your products under your original quotation. If you raise the prices this year, it will be more difficult for us to compete others' in this market.

As you know, the devaluation of US$ to NT$ was 3~5% in the past three months, besides, the local wages and overhead cost have been much increased. In fact, we almost have no space in selling your products here.

Please try your best to make the cost down in every respect and let us have your revised quotation by end of this month.

Best wishes,

Jack Turner
Purchasing Manager

翻譯

請注意，今年在您原始報價下推廣您的產品是非常困難的，如果您在今年還要提高價格，我們將會更困難與其他競爭者競爭。

如您所知，過去三個月美元對新台幣的貶值為 3~5%。此外，當地的工資和製造費用大幅增加，實際上在這裡銷售您的產品我們幾乎沒有利潤。

請盡可能降低各方面的成本，並於本月底之前修改報價給我們。

範例 6-6　買方：客戶還價，並詢問有無較便宜的替代品

To	lindachen@tbi.com.tw
From	jackturner@pioneer.com
Cc	
Bcc	
Subject	Your offer for cell phone case
Attachment	

Dear Linda,

Your quotation for cell phone case is FOB US$16/case which is not competitive enough. Our target price is US$12/case. We suggest you re-examine the cost, also study the alternative materials or production method to reduce the cost. If you cannot reduce the price, is there any substitute can meet our requirement?

Please reply the above before this Friday.

Best wishes,

Jack Turner
Purchasing Manager

翻譯

您的手機殼報價為 FOB 16 美元／件，這樣的價格是沒有競爭力的。我們的目標價格是 12 美元／件。我們建議您重新檢查成本，並研究替代材料或生產方法以降低成本。如果您不能降低價格，有沒有替代品可以滿足我們的要求？

請在本週五之前回覆我們。

範例 6-7　買方：客戶的訂購量大，請賣方降價

To	lindachen@tbi.com.tw
From	jackturner@pioneer.com
Cc	
Bcc	
Subject	Your Quotation No. Q5612
Attachment	

Dear Linda,

We thank you for your quotation dated on 3/15. While appreciating the good quality of your sweaters, we find the prices rather high for our customers to accept. In addition, we have to point out that very good sweaters are now obtainable here from several China manufacturers, and all of their prices are from 10% to 15% below yours.

However, we would like to place our order with you but must ask you consider whether you can make us a more favorable offer. As far as we can see, our order will be worth around US$300,000, you may think it worthwhile to make a concession.

We are looking forward to your reply by end of this month.

Best wishes,

Jack Turner
Purchasing Manager

翻譯

感謝您在 3 月 15 日的報價。我們很滿意您毛衣的品質，但是我們發現您的報價相當高，這樣我們的客戶難以接受。另外，我們必須指出，現在很多中國製造商都可以提供非常好的毛衣，而他們的價格都比你低 10%到 15%。

雖然如此，我們還是希望向您下訂單，但必須要求您考慮是否可以在報價上再優惠些。我們目前的訂單預估量約為 30 萬美元，我相信這樣的訂單量可以讓您在價格上再考慮一番。

我們期待您能在本月底回覆我們。

範例 6-8　賣方：答應降價

To: jackturner@pioneer.com
From: lindachen@tbi.com.tw
Cc: judywong@ttm.com.tw
Bcc:
Subject: Price for Beach Sandal
Attachment:

Dear Jack,

Thank you very much for your e-mail, in which you ask for a reduction of our quotation. In consideration of the start business with your company, we have decided to accept your proposal.

We hope that this concession will result in a considerable increase in your orders and assure you that we will always do our utmost to execute them to your complete satisfaction.

Best regards,

Linda Chen
Sales Manager

翻譯

非常感謝您的來函，信函中您希望我方的價格要再降低。我方考慮到能與貴公司開始發展雙方業務，我們決定接受您的要求。

我們希望這一次的降價可以使您的訂單大幅增加，並向您保證，我們將始終竭盡全力執行您的訂單以使您完全滿意。

範例 6-9　賣方：如果增加數量則同意降價

To jackturner@pioneer.com
From lindachen@tbi.com.tw
Cc judywong@ttm.com.tw
Bcc
Subject Price for Beach Sandal
Attachment

Dear Jack,

We are very sorry to learn from your e-mail that you were not able to accept our offer. We have to point out that though our goods are slightly higher in price than similar articles, their quality is far superior to those of other manufacturers as shown by the orders we have been receiving recently.

However, in order to show our sincere interest in your business, we will meet your request by granting you a special discount of 5% if you can increase your order to 10,000 pairs per year.

We look forward to your reply.

Best regards,

Linda Chen
Sales Manager

翻譯

我們很抱歉從您的電子郵件中了解到您無法接受我們的報價。我們必須指出，儘管我們的商品價格略高於其他同類商品，但它們的質量遠遠優於其他製造商的質量，正如我們最近收到的訂單所顯示的那樣。

然而，為了表示我們對您業務的真誠重視，我們將接受的要求，如果您每年可以將訂單增加至 10,000 雙，我們就可以給予您 5%的特別折扣。

我們期待您的回覆。

範例 6-10　賣方：價格已是最低，無法再降

To: jackturner@pioneer.com
From: lindachen@tbi.com.tw
Cc: judywong@ttm.com.tw
Bcc:
Subject: Price for Beach Sandal
Attachment:

Dear Jack,

We thank you for your counter offer on Aug. 5. We certainly would like to support you regarding the matter of price. However, we have to inform you that there is no room for further reduction in prices as we have already cut them to the absolute minimum.

We await your acceptance of this offer.

Best regards,

Linda Chen
Sales Manager

翻譯

我們感謝您在8月5日來函的還價。我們當然希望在價格上為您提供最好的價格。但是，我們必須告知您，由於我們已經將價格降至最低，所以價格無法有進一步下調的空間。

我們希望您能接受此報價。

範例 6-11　賣方：價格已是最低，無法符合對方價格要求

To: jackturner@pioneer.com
From: lindachen@tbi.com.tw
Cc: judywong@ttm.com.tw
Bcc:
Subject: Price for Cell Phone Case
Attachment:

Dear Jack,

We regret to hear that you can not accept the price we offered. After we carefully checked your target price, we are really sorry to inform you that US$12/case is under our cost. To avoid the quality problem happened after shipment, we will not consider to adopt any other alternative.

Although this time we may not reach your target price, we hope we can cooperate with you in the future deals.

Best regards,

Linda Chen
Sales Manager

翻譯

我們很遺憾聽到您不能接受我們提供的報價。經我們仔細檢查了您的目標價格後，我們非常抱歉地通知您，您的目標價 12 美元／箱，這樣的價格是低於我們的成本。為避免發生出貨後才產生的品質問題，我們不會考慮採用其他替代方式來降低品質而降低價格。

雖然這次我們可能無法達到您的目標價格，但我們希望我們能夠在未來的業務上與您合作。

範例 6-12　賣方：利潤有限價格已是最低，另推薦替代品

To	jackturner@pioneer.com
From	lindachen@tbi.com.tw
Cc	judywong@ttm.com.tw
Bcc	
Subject	Price for P/N: AA213B
Attachment	

Dear Jack,

We have received your counter offer. To our regret, we are unable to offer you any further discount. Refer to our quotation on 6/18, we already offered you our lowest price. Frankly speaking, production cost has risen since March, and our profit margin doesn't allow us any concession by offering any price cut or discount.

However, in order to meet your requirement, we would like to recommend our model BB215H as an excellent substitute for the goods you need. This is superior in quality and we will offer a special discount of 3% off our list price only for this quarter. If you are interested in this model, please do take it into consideration and confirm as early as possible.

Best regards,

Linda Chen
Sales Manager

翻譯

我們已收到您的還價。遺憾的是，我們無法再提供更多折扣。請參閱我們在 6/18 的報價，我們已經報了我們最低的價格。自 3 月份以來製造成本一直上漲，我們的利潤已經非常低，所以我們無法再降價或是提供任何折扣。

但是，為了滿足您的要求，我們希望推薦我們的 BB215H 型號作為替代品，這款商品可以符合您的需求。這一款的品質優良，而且我們可以給您 3% 的特別折扣，但是這一折扣只在本季有效。 如果您對此型號感興趣，請考慮並儘早確認訂單。

常用句

買方

1. We thank you for your offer, but our prices are too high.
2. In view of the prevailing prices in this market, your quotation is not competitive.
3. The prevailing prices in this market are 10% higher than your limit price.
4. A comparison of your price with that of our other suppliers shows that their quotation is more attractive.
5. We hope you will realize the importance of a further shading of the prices.
6. Your price is too high, and competitors' prices are more attractive.
7. It would not be possible for us to purchase the quantity at the price you offer.
8. Will you reduce the prices of the articles mentioned below?
9. We may give an order if you can reduce the price to some extent.
10. This order is quite large, and we would request that you make a quantity discount.
11. We can only consider placing an order if you can give a price reduction of 5%.
12. We will work harder to secure orders if you support us with lower prices.

賣方

1. In view of the long business relations with your company, we accept the terms you specified for this order.
2. In view of the long business relations, we are compelled to accept your demand for discount.
3. We can offer you a discount of 2% from the list price.
4. Under consideration, we will offer you an allowance of 5% off the list price.
5. In reply to your request, we are willing to allow you an extra discount of 3% on this order.

6. We agree to make a reduction in price if this will help you to develop your market for our products.
7. We can grant you a 5% discount on orders exceeding US$10,000 in value.
8. The quantity discounts vary according to the volume of the order.
9. The prices have been rising steadily, but we will make a special discount for you.
10. We regret that your price does not allow us to accept the order.
11. We regret to learn that you find our quotation rather expensive.
12. We could not be in a position to accept your proposal.
13. Please kindly understand that we have quoted our rock-bottom prices. Your price is indeed unacceptable to us.
14. The price you specified is too low to tempt us to make a counter offer.
15. We have to decline the order at the price you stipulated.
16. Our margin of profit is so small that is is impossible for us to make any further concession.
17. Wages and raw materials have risen substantially, and we are compelled to adjust our prices to cover at least part of this rise.
18. As this is the case, we are not in a position to make any further discount.
19. Our price does not allow any further discount.
20. Under the circumstances, we could not grant you a reduction.

練習題

1. 將下列的句子翻成英文

A. 根據目前的市價，您報的價格偏高了一點。

B. 您的報價與其他供應商的價格相較之下，他們的價格較優惠。

C. 謝謝您的答覆，不過您的報價偏高，無法符合我們的市場。

D. 我們無法以此種價格購買這樣的數量。

E. 這一張訂單的數量極大，希望您能因量降價。

F. 如果您願意打九折，那我們可以考慮下單。

G. 如果您能以較低的價格來支持我們，我們將更努力去客戶那裡爭取訂單。

H. 由於生意一質不佳，希望您可以改變一下付款條件。

I. 如果您能符合我們的條件，我們就會下訂單。

J. 由於付款上有些不方便，希望您能將及期票延長成 90 天。

K. 基於長期合作生意的觀點，我們接受您針對此次訂單所提出的條件。

L. 為了能打入貴國的市場，我們願意減少利潤，降低 4% 的價格。

M. 如果降價可以打入您們的市場，我方願意降價。

N. 只要數量超過 1000 台以上，我方就可以再額外打九五折給您。

O. 我們的目標價是每個美金 10 元，請告知貴方是否可接受。

P. 你們的價格比同行貴 5%。

2. 將下列中文書信翻譯成英文

感謝您 3/2 來函的報價單與毛衣的最新型錄，但是我們發現你們的報價偏高。

我們要訂的數量是 10,000 打，請重新報較好的價格給我們。我們的目標價格是每打美金 65 美元，如果貴方可以接受，請確認，並提供一件免費的樣品以供品質確認。

我們靜候貴公司的回音。

3. 買方：寫一封還價信，告知賣方報價太高，無法與市場上的其他競爭者競爭，請考慮再降價 5% 我方才有興趣下單。

4. 賣方：回覆客戶還價，表明已報最好的價格，無法再降價 5%，且買方的數量少產品多，實在沒有任何降價的空間，請對方了解。如果對方未來可以將數量增加到一定數量，我方可以再降價。

5. 賣方：為表示對客戶的支持，我方將於這一張訂單考慮接受客戶的目標價格：FOB Taiwan US$3.50/pc，但是訂單需於一星期內讓我方收到，並且我方只提供標準顏色。

6. 賣方：因近日的原物料及工資上漲，我方將於明年度調漲價格，今年的價格將盡力維持原價，敬請及早下單，以舊價取得商品。

CHAPTER

07 樣品與規格

Sample & Specification

買方通常在賣方報價或是自己自動詢價之後，對於有興趣的商品會進一步索取賣方的標準樣品。有時候買方會有自己的規格，需賣方進一步根據買方的需求去做樣品，這樣的過程稱為「打樣」。有時候打樣的過程會牽涉到需不需再開模具、模具費買方如何負擔等問題，這一些問題都需買賣雙方進一步討論並同意後，賣方才會開模具再進一步打出買方規格的樣品。

買方收到樣品後，如果符合規格要求，買方需寄樣品規格確認的信函給賣方。賣方是否要索取樣品費與運費則由賣方決定。通常對客戶索取樣品的處理方式有下列幾種情況：

1. 樣品和郵費全部免費 (Both sample and postage are free.)
2. 樣品免費，但是客戶需支付郵費，因為如果樣品的重量很重或是材積很大，則運費會很高 (The sample is free, but you have to pay for the postage because shipping charge is expensive.)
3. 樣品和郵費都需事先支付，因為材料成本高 (Please pay sample charge and postage in advance because the material cost is high.)
4. 請客戶先支付樣品費和郵費，等客戶正式下單時，再退還或是從訂單中扣除。(Please pay the sample charge and postage in advance. However, we will return them to you when ordering, or deduct them from the order.)

樣品一經確認，買方即可下訂單於賣方，賣方出貨的商品品質需與樣品一致。賣方在送樣確認時須注意：

1. 大量量產時，商品品質與樣品一致。如果於生產中或是出貨前，發現商品與樣品有所不同時，一定要在發現時或出貨前告知買方，如買方同意不同處，請對方發確認函後再行出貨。或是重送樣品請買方重新確認，否則不但有遭退貨的危險，還可能需賠償買方損失。
2. 樣品確認前，不可出貨。
3. 有時客戶要的尺寸、規格，賣方目前正好沒有，可以先送類似的樣品，供買方先作品質的參考，等買方正式訂購樣品時，再送正確的樣品供正式的確認。

範例 7-1　買方：接受報價，索取免費樣品

To: lindachen@tbi.com.tw
From: jackturner@pioneer.com
Cc:
Bcc:
Subject: Price Confirm & Sample Request
Attachment:

Dear Linda,

We confirm the acceptance of your price quoted in your quotation of April 1. However, before placing the order, please send us 3 free samples for your model no. AA567H.

We will place you our formal order upon sample approval. Hope to receive the samples within one week, please speed them up.

Best wishes,

Jack Turner
Purchasing Manager

翻譯

我們確認接受您 4 月 1 日的報價。但是，在下訂單之前，請寄給們 3 份免費樣品，型號 :AA567H。

樣品確認後，我們會將會向您下正式的訂單。希望在一周內收到樣品，請儘快。

範例 7-2　賣方：回覆客戶並寄送樣品

To	jackturner@pioneer.com
From	lindachen@tbi.com.tw
Cc	
Bcc	
Subject	Sample for model no. AA567H
Attachment	

Dear Jack,

Thanks for your e-mail of May 23 confirming your acceptance of our prices and requesting samples.

We will send you three samples of model no. AA567H by express air parcel post for your quality approval tomorrow. Please let me know if you have any further questions.

Best regards,

Linda Chen
Sales Manager

翻譯

感謝您 5 月 23 日的來函，確認接受我們的報價並索取樣品。

我們會寄給您 3 個 AA567H 的樣品，樣品將於明天由航空包裹寄給您，以利您的的品質審查。如果您還有其他問題，請通知我。

範例 7-3　買方：發現樣品少了，請賣方補寄

To lindachen@tbi.com.tw

From jackturner@pioneer.com

Cc

Bcc

Subject Shortage of Samples

Attachment

Dear Linda,

I received your samples today, however, I did not find the samples for model BB-532 and the relevant specifications inside. We do need this model urgently. Please airmail them immediately so that I can receive the samples by this Friday.

Best wishes,

Jack Turner
Purchasing Manager

翻譯

我今天收到您的樣品了。但是，我沒有發現型號 BB-532 的樣品和相關規格。我們迫切需要這個型號的商品。煩請立即空運這一個商品的樣品給我們，以便我可以在本週五之前收到樣品。

範例 7-4　賣方：回覆客戶並補寄樣品

To	jackturner@pioneer.com
From	lindachen@tbi.com.tw
Cc	
Bcc	
Subject	Sample for model no. BB-532
Attachment	

Dear Jack,

We are sorry for our neglect that the samples sent to you with shortage of BB-532. We have sent 3 samples of BB-532 to you by air parcel today. You can receive them by end of this week.

If you have any further questions, please don't hesitate to contact me via e-mail or call me at 1-886-22578-8888 ext. 456.

Best regards,

Linda Chen
Sales Manager

翻譯

我們很遺憾寄送的樣品中竟然少了 BB-532 的樣品。我們今天用航空包裹寄了 3 個 BB-532 樣本給您。你可以在本週末收到。

如果您還有其他問題，請隨時通過電子郵件與我聯繫，或撥打我的電話 1-886-22578-8888 分機 456。

範例 7-5　買方：同意賣方開模，並請先寄樣品

To lindachen@tbi.com.tw

From jackturner@pioneer.com

Cc

Bcc

Subject Samples of AA567H

Attachment

Dear Linda,

We accept your quotation for model AA567H including mold cost. I know you will spend 2 months to make the mold. However, before starting the mold, we would like to know if you can make a handmade sample for our test first. We also can look at the appearance to see if there is anything needs further modification. If there is no big issue, we can place mold order right away.

Best wishes,

Jack Turner
Purchasing Manager

翻譯

我們接受您對 AA567H 型號的報價，其中包括模具成本。我知道你們需 2 個月來製作模具。但是，在製作模具之前，我們想知道您是否可以先以手工製作樣品讓我們可以先測試一下。這樣我們可以先看看外觀，看看是否有什麼需要做進一步修改。如果沒有什麼大問題，我們就可以立即下模具訂單。

範例 7-6　賣方：回覆客戶手製樣品需收樣品費

To jackturner@pioneer.com

From lindachen@tbi.com.tw

Cc

Bcc

Subject Samples of AA567H

Attachment

Dear Jack,

Thank you for your acceptance of our quotation for model AA567H. It is no problem to make a handmade sample for your test first. However, since the sample charge for the handmade sample is very high we have to ask you to pay for the sample charge US$80/pc. The lead time for sample will take one week.

Please confirm then we can proceed samples.

Best regards,

Linda Chen
Sales Manager

翻譯

感謝您接受我們對 AA567H 型號的報價。首先為您先以手工製作樣品一個測試樣品是沒有問題的。但是，由於手工樣品的樣品收費非常高，我們必須要求您支付樣品費 80 美元／件。樣品的交貨時間將需要一周的時間。

請確認，然後我們可以進行樣品試作。

範例 7-7　買方：樣品測試未通過，請賣方整理後再送

To: lindachen@tbi.com.tw
From: jackturner@pioneer.com
Cc:
Bcc:
Subject: Samples of AA567H
Attachment: test report

Dear Linda,

Enclosed is the test report of your model no. AA567H from our customer. Please check and correct the deviations. Also advise how long you need for the reworking and re-sending samples for approval.

We are waiting for your reply immediately.

Best wishes,

Jack Turner
Purchasing Manager

翻譯

附件是你的型號 AA567H 的測試報告，這一個報告來自我們的客戶。請詳細檢查此報告並糾正產品的偏差。同時告知您需要多長時間來重新製作樣品並重新發送樣品以供審核。

期待您儘速回覆。

範例 7-8　賣方：回覆客戶會研究測試報告後再寄樣品

To	jackturner@pioneer.com
From	lindachen@tbi.com.tw
Cc	
Bcc	
Subject	Re:Sample for model no. AA567H
Attachment	

Dear Jack,

I have received test report for model no. AA567H. I will forward the report to our engineer for further study immediately. I think the deviation should be a minor issue and it will not cost much time to correct the samples. Let me check and give you the exact time for sending new samples tomorrow.

Best regards,

Linda Chen
Sales Manager

翻譯

我收到了型號 AA567H 的測試報告。我會立即將報告轉發給我們的工程師作進一步研究。我認為這種偏差應該是一個小問題，並且不會花費太多時間來糾正樣本。我先檢查一下，明天再告訴您送新樣本的確切時間。

範例 7-9　買方：去函索樣

To	lindachen@tbi.com.tw
From	jackturner@pioneer.com
Cc	
Bcc	
Subject	Samples of LM-123
Attachment	

Dear Linda,

We have received your quotation and latest catalog. We are interested in your products especially the item LM-123. We would like to ask one sample of this item for evaluation. Please send us one sample before 23rd this month and the sample charge will be paid via T/T upon receipt.

Best wishes,

Jack Turner
Purchasing Manager

翻譯

我們已收到您的報價和最新目錄。我們對您的產品特別是 LM-123 感到興趣。請您寄一個這個商品的樣品給我們以供評估。請在本月 23 日前寄給我們一份樣品，樣品費我們將於收到樣品後以電匯的方式支付。

範例 7-10　賣方：回覆客戶目前沒有現貨樣品，可否寄替代品

To: jackturner@pioneer.com
From: lindachen@tbi.com.tw
Cc:
Bcc:
Subject: Re:Sample for model no. LM-123
Attachment:

Dear Jack,

Thanks for your e-mail in requesting a sample of LM-123. However, we do not have this item in stock right now. This sample will be available in July after our plant schedules production next month. You will not be willing to wait so long to receive the sample, therefore, I would suggest you to take the similar item (P/N:LM-169) as a substitute for quality evaluation as both quality and functions are the same except there is a little difference in appearance.

Please confirm if you would like me to send you the sample of LM-169.

Best regards,

Linda Chen
Sales Manager

翻譯

感謝您來函所取 LM-123 樣品。但是，我們現在沒有這個產品的庫存。 這個樣品將在 7 月才有可能提供，因為我們工廠會在下個月才生產這個型號的商品。我們想您應該不會願意等待這麼長的時間。因此，我建議您可以考慮相似的商品（型號：LM-169）來評估品質與替代性，因為這兩款的品質和功能都是相同的，除外觀有點不同。

請確認您是否希望我寄給您 LM-169 的樣本。

常用句

買方

1. We acceptance of your prices quoted in your quotation of June 3rd.
2. We are interested in your products, especially model no. AAA, please send 3 samples for evaluation.
3. We received your samples and these samples have deviation shown on attached report. Please make a correction and re-send.
4. Sample charge will be paid immediately upon receipt.

賣方

1. In reply to your inquiry of May 10, we are pleased to send you samples.
2. We are sending you our catalogue together with samples under separate cover.
3. We are glad to inform you that we sent you our samples via air parcel today.
4. The sample you request is not in stock now, and it will not be available in 3 months.

練習題

1. 將下列的句子翻成英文

A. 很高興收到您 5 月 12 日的來函，通知您我們已經將您要的樣品寄去給您了。

B. 我們以分別寄出本公司的目錄及樣品給您。

C. 如您所指示的，我們將最新的目錄寄給您，其中包含了各式產品的插圖和規格說明。

D. 如您所指示的寄上手機的樣品、目錄及價格表。

E. 很高興另郵寄上我們的報價單及樣品給您。

F. 我們將下給你一張樣品訂單。

G. 下訂單以前，我們希望收到你們的樣品。

H. 如所要求，我們用郵包寄上一個免費樣品以供確認品質。

I. 我們希望很快收到您的樣品訂單。

2. 將下列中文書信翻譯成英文

我們已收到您的報價單 Q2017158，但是下單以前，可否請您各寄上 2 個免費樣品以供我們作進一步的評估。樣品一經確認，我們會馬上下正式訂單給你們。

我們希望於一週內收到，請加速送來。

3. 賣方：就以上來函，回一封信。

4. 請寫一封索取手錶樣品的索樣信函。

5. 根據第 4 題的信函回覆客戶，因手錶價格高，因此要索取樣品費 US$150，但若是客戶日後下單，樣品費會從訂單中扣除。

6. 寫一封信函通知客戶對其索求的樣品可以免費提供，但是客戶需付運費

7. 樣品寄出一週後，請寫一封 e-mail，跟催客戶。

CHAPTER

08 追蹤

Follow Up

在各種信函發出後，如果在一定期間內沒有收到對方的回應，我們就應該主動發函去追蹤。追蹤信包含在對新客戶發出推銷信、詢問信、報價信、客戶來函索樣後的後續追蹤。除此之外，對新客戶第一次下單的追蹤與付款的追蹤也是重要的一環。

除了推銷信的追蹤以及很久沒聯絡的客戶的追蹤，在信函上需要一些技巧與禮貌之外，其他一般信函的追蹤信寫法都很直接。如果是對客戶訂單的追蹤，在跟催的過程中，可以藉由各項理由與說詞來加強跟催信函的內容，例如：

- 市場需求增加
- 大量訂單湧入
- 匯率波動大，但維持原價至年底
- 生產線吃緊
- 購買量大，願給額外的折扣
- 商品品質佳，價格具競爭力

範例 8-1　賣方：參展後，追蹤來參觀過的客戶 I

To	jackturner@pioneer.com
From	lindachen@tbi.com.tw
Cc	
Bcc	
Subject	Trade Show of Chicago International Fair
Attachment	

Dear Jack,

First of all, we would like to thank you for your visit our booth and showing interest in our products at the Chicago International Fair last month.

As inquired, we are glad to send you our catalog of standing fan for your evaluation. Please specify the items interest you, or send us your detailed specification for making exact quotation.

We look forward to building mutually beneficial relationship with you soon.

Best regards,

Linda Chen
Sales Manager

翻譯

首先，我們非常感謝您在上個月的芝加哥國際商展上來參觀我們的展位，並對我們的產品有興趣。

依您所詢問，我們很高興寄給您我們的立扇型錄以供評估。請告知我方您於型錄上感興趣的商品，或將您想要的商品的詳細規格給我們，以利進行確切的報價。

我們期待儘快與您建立互惠關係。

範例 8-2　賣方：參展後，追蹤來參觀過的客戶 II

To	jackturner@pioneer.com
From	lindachen@tbi.com.tw
Cc	
Bcc	
Subject	Trade Show of Chicago International Fair
Attachment	Quotation

Dear Jack,

Please refer to our e-mail dated Nov. 16 and our quotation Q7789. Would you please tell us of your comments or order information?

From the samples we showed you at the fair, you have seen that in view of the fine quality and attractive designs. Our goods are really of good quality for the price. We can assure you that these goods have a great success in Western markets, in which we have had very successful experience. If you can confirm an order within 2 weeks (before Nov. 30) and place an order (quantity over 10,000 pcs), we can grant you a special discount of 2%.

Please kindly take this good opportunity by confirming return soon.

Best regards,

Linda Chen
Sales Manager

翻譯

我們於 11 月 16 日的電子郵件上有附上我們的報價單 Q7789。想請問您對報價單是否有任何的意見或是有任何訂單信息嗎？

從我們在展會上向您展示的樣品中，您可以看到，我們的商品有優良的品質和吸引人的設計。我們的商品品質以這種價格銷售真的是很好的選擇。我們可以向您保證，這些商品在西方市場上銷售良好，我們在這方面擁有非常成功的經驗。如果您可以在 2 週內（11 月 30 日前）確認訂單並下單（數量超過 10,000 個），我們可以給您 2%的特別折扣。

請儘快確認並回覆來獲取這個好機會。

範例 8-3　賣方：對新開發客戶的追蹤

To jackturner@pioneer.com

From lindachen@tbi.com.tw

Cc

Bcc

Subject Our latest catalog

Attachment

Dear Jack,

I sent you our latest catalog on Aug. 15 together with our best quotation. We are wondering if you have well received them since I have not received any response from you till now.

From our catalog and the quotation enclosed, you will find that we have many innovation products with the best quality at the competitive prices for customers. Please don't hesitate to let me know your interest and comments soon.

Best regards,

Linda Chen
Sales Manager

翻譯

我於 8 月 15 日寄給您我們最新的產品目錄以及我們最好的報價。我們想知道您是否已經收到，因為我至今尚未收到您的回覆。

從我們的產品目錄和附件的報價中，您會發現我們有許多創新產品，這些產品我們以最具競爭的價格為客戶提供最優質的產品。請不要猶豫告知我們您有興趣的產品和意見。

範例 8-4　賣方：拜訪客戶之後的追蹤

To jackturner@pioneer.com
From lindachen@tbi.com.tw
Cc
Bcc
Subject Meeting Minutes of June 18
Attachment cost details.doc

Dear Jack,

Firstly I would like to thank you for your kind hospitality during our visit. It was great to visit your office finally.

According to our meeting on June 18, attached please find the breakdown cost about the mold charge and freight cost for products. If you need any further information, please let me know, and I look forward to developing a long term business relationship with your company.

Best regards,

Linda Chen
Sales Manager

翻譯

首先我要感謝您在我們拜訪期間的熱情款待。最後能夠拜訪您的辦公室真是太棒了。根據我們 6 月 18 日的會議，附件請查看關於產品的成本分析表、模具費用和運費。如果您需要更多資料，請不吝告知，我方期待與貴公司建立長期業務關係。

範例 8-5　賣方：寄給客戶推銷信後的追蹤 I

To: jackturner@pioneer.com
From: lindachen@tbi.com.tw
Cc:
Bcc:
Subject: New Phone Case
Attachment:

Dear Jack,

It is our pleasure to introduce new smart phone case model ST741 to the respective customer like you continually. We believe that you should have received our catalog sent on Feb. 25.

For your information, we are opening the new mold for this new model which will be finished by the end of this month. Accordingly, we will prepare the relative samples for your evaluation in case you are interested in it. Persisting the principles of superiority in quality, innovation in products, and integrity in business, we have won a very good reputation in the world market.

We believe to cooperate with us will bring considerable profit for your business. Please let us know of your interest and comments soon.

Best regards,

Linda Chen
Sales Manager

翻譯

我們很高興向您這樣優質的客戶介紹我們不斷推出的新智慧型手機機殼，型號 ST741。我們相信您應該有收到我們在 2 月 25 日寄出的目錄。

在此告知您，我們正在為這新款式開發新的模具，該模具將於本月底完成。因此，如果您對此商品有興趣，我們會提供相關樣品供您評估。我們堅持卓越的品質，產品的創新和誠信經營的原則，在國際市場上贏得了良好的聲譽。

我們相信與我們合作將為您的業務帶來可觀的利潤。請儘快告訴我們您對此商品的興趣和意見。

範例 8-6 賣方：寄給客戶推銷信後的追蹤 II

To jackturner@pioneer.com
From lindachen@tbi.com.tw
Cc
Bcc
Subject Modern system of lateral filing
Attachment

Dear Jack,

As we have not heard from you since we sent you our catalog of filing systems, I wonder whether you require any further information before placing an order.

The modern system of lateral filing has important space-saving advantages whether economy of space is important. However, if space is not a challenge for you, our flat-top suspended system may be suitable. This system gives a neat and tiny appearance to the filing drawers, and files can be located easily and quickly. Many customers find these two features particularly beneficial.

Our representative, Richard Robison will visit US next month. Would you like us to arrange a visit to your office? Richard has advised on equipment for many large, modern offices. He would be able to recommend the system most suitable for your market. There would of course be no obligation. Alternatively, you may prefer to visit our show room and see for yourself how the different filing systems work.

I look forward to hearing from you soon, or you may contact me on 1-886-22567-884 with any questions.

Best regards,

Linda Chen
Sales Manager

翻譯

我們寄給您我們的檔案櫃目錄後一直尚未收到您的回信，因此我想知道您在下訂單之前是否需要更多資料。

我們最新款的側邊邊櫃對於空間的節省在現代的空間設計上有相當的優勢。但是，如果空間不是您最重要的考量，我們上方平面的懸掛系統櫃可能更合適。這個系統給文件抽屜提供了一個整潔和小巧的外觀，文件可以很容易和快速地找到。許多客戶發現這兩個功能特別有用。

我們的業務代表 Richard Robison 將於下個月訪問美國。我們可以安排他去拜訪您嗎？Richard 曾為許多大型現代化辦公室提供設備諮詢。他將能夠推薦最適合您市場銷售的系統櫃。當然我們是非常樂意提供建議的。或者，您可以拜訪我們的展示廳並親自了解不同的文件系統是如何作業的。

我期待能儘快收到您的回覆，如果您有任何問題歡迎與我聯繫，電話：1-886-22567-884。

範例 8-7　賣方：寄給客戶報價與樣品後的追蹤（單價下降，請客戶下單）

To	jackturner@pioneer.com
From	lindachen@tbi.com.tw
Cc	
Bcc	
Subject	Price for LED products
Attachment	

Dear Jack,

Please kindly refer to our e-mail dated Sep. 18, do you have any conclusion now? Would you please confirm a trial order soon?

For starting our favorable cooperation with you, we have promised you to lower unit price from US$3.50 to US$3.25 for the trial order. Frankly speaking, our profit is very limited so it is indeed the best we can offer. Please kindly understand. Your prompt confirmation will be appreciated.

We look forward to hearing from you soon.

Best regards,

Linda Chen
Sales Manager

翻譯

我們於 9 月 18 日寄給您的信函，請問您目前有任何結論了嗎？您能否儘快確認試用訂單？

為了開始我們雙方的業務合作，我們承諾您將單價從3.50美元降至3.25美元。坦白說，我們的利潤非常有限，所以它確實是我們能提供的最好的價格，請體諒。您立即的確認將不勝感激。

我們期待您的佳音。

範例 8-8　賣方：寄給客戶報價與樣品後的追蹤（已報價，請客戶下單）

To jackturner@pioneer.com
From lindachen@tbi.com.tw
Cc
Bcc
Subject Price for LED products
Attachment

Dear Jack,

Regarding our quotation Q12546 and sample sent to you on Mar. 20, do you have any conclusion now? If yes, please kindly inform us of your order contents in details so that we can reserve a space for you accordingly.

We look forward to your order confirmation soon.

Best regards,

Linda Chen
Sales Manager

翻譯

關於我們在 3 月 20 日寄給你的報價單 Q12546 和樣品，請問您目前有什麼結論嗎？如果有，請詳細告知我們您的訂單內容，以便我們可以為您預留相應的生產線。

我們期待您的訂單確認。

範例 8-9 賣方：寄給客戶報價與樣品後的追蹤（價格上漲，請客戶及早下單）

To jackturner@pioneer.com
From lindachen@tbi.com.tw
Cc
Bcc
Subject Price for LED products
Attachment

Dear Jack,

Refer to our e-mail of May 30 in which we have offered you our best prices, would you please confirm your order as soon as possible?

As raw materials and wage have increased a lot for past few months, for supporting you, we will maintain our offer until July 5. Please kindly understand our sincerity and confirm your order as early as possible in order to avoid further rises in costs.

Best regards,

Linda Chen
Sales Manager

翻譯

請參閱我們於 5 月 30 日寄給您的電子郵件，信中我們提供了最優惠的價格，可以請您儘快確認您的訂單嗎？

由於原材料和工資在過去幾個月增加了很多，為了支持你的業務發展，我們將保留我們的報價有效期至 7 月 5 日。請了解我們的誠意，並儘早確認你的訂單，以避免成本進一步上漲。

範例 8-10　賣方：寄給舊客戶的追蹤 I（生產線忙，請客戶及早下單）

To jackturner@pioneer.com

From lindachen@tbi.com.tw

Cc

Bcc

Subject Price for LED products

Attachment

Dear Jack,

Please refer to our quotation for LED products. Are you going to place a repeat order recently?

In the near future we will arrange a big order on our production line which may occupy about 2 months. Therefore, we would like to ask for your purchase forecast first and if needed, we can arrange your order first to avoid any delay of your repeat order.

Please kindly take this as an urgent matter and confirm return ASAP.

Best regards,

Linda Chen
Sales Manager

翻譯

請參閱我們 LED 產品的報價。請問您最近要持續下單嗎？

因為在不久的將來，我們將在我們的生產線上安排一個可能需要約 2 個月生產的大訂單。因此，我們想請您先提供您的採購數量預測，如果需要，我們可以先安排您的訂單生產，以避免延誤您的訂單。

請將此作為緊急事項考量並儘快與我們確認。

範例 8-11 賣方：寄給舊客戶的追蹤 II（很久沒下單）

To jackturner@pioneer.com
From lindachen@tbi.com.tw
Cc
Bcc
Subject New Product – Beach Sandal
Attachment

Dear Jack,

It has been a long time we have not contacted with you. We sincerely wish you having a prosperous business.

Attached please find our latest catalog for your reference in which we have launched a new beach sandal recently. From our past communication, we know of your interest in our subject products. Please take a look into it and feel free to contact us if this item interests you.

We will accordingly offer you the best quotation and samples right away. Thank you for your kind attention and look forward to renewing and continuing the good business relationship with you very soon.

Best regards,

Linda Chen
Sales Manager

翻譯

我們已經有很長一段時間沒有與您聯繫。我們真誠希望您的業務一直欣欣向榮。

在此信函內我們請附上最新的產品目錄，供您參考。我們最近推出了一款新的沙灘涼鞋。從我們過往的聯繫中，我們了解您會對我們哪一種商品感興趣。請看看這一款新的沙灘鞋，如果您對此商品感興趣，請隨時與我們聯繫。

我們會立刻提供最好的報價和樣品給您。感謝您的關注，並期待我們雙方能儘快更新並繼續保持良好的業務關係。

範例 8-12　賣方：寄給舊客戶的追蹤 III（很久沒下單）

To	jackturner@pioneer.com
From	lindachen@tbi.com.tw
Cc	
Bcc	
Subject	New sports goods
Attachment	catalog

Dear Jack,

We notice that it is some time since we last received an order from you. We hope this is in no way due to dissatisfaction with our service or with the quality of goods. In either case we would like to hear from you.

We are most anxious to ensure that customers obtain maximum satisfaction from their dealings with us. If lack of orders from you is due to changes in the type of goods you handle, we may still be able to meet your needs if you will let us know in what directions your policy has changed.

As we have not heard otherwise, we assume that you are still selling the same range of sports goods, so a copy of our latest illustrated catalog is enclosed. We feel this compares favorably in range, quality and price with the catalog of other manufacturers. You will see in our catalog that our terms are now much better than previously, following the withdrawal of exchange control and other official measures since we last did business together.

I hope to hear from you soon.

Best regards,

Linda Chen
Sales Manager

翻譯

我們注意到距我們上次收到您的訂單已有一段時間了。我們希望這絕不是因為您對我們的服務或商品質量的不滿。無論哪種情況，我們都希望能收到您的來信。

我們希望確保客戶從與我們的交易中能獲得最大滿意度。如果您沒有對我們下訂單是由於您所處理的商品類型發生變化而導致的，我們還是可以提供滿足您需求的服務，如果您能讓我們知道您目前處理的商品與購買的政策。

因為我們沒有聽到任何改變的情況，我們假設您仍然在銷售同樣的體育用品，因此附上了我們最新的目錄。我們認為比較我們的型錄與與其他製造商的產品目錄，您可以發現我們的商品在商品類別，品質和價格方面都優於其他供應商。在我們的目錄中，您也可以發現因為外匯管制撤銷和其他官方措施，我們的付款條款也比以前好的。

我希望能儘快獲得您的消息。

常用句

1. We have not heard from you since we sent you our catalog last month.
2. We have not received any comments from you since we sent you quotation last month.
3. We strongly advise you to avail yourselves of this exceptional opportunity.
4. We advise you to accept this offer without loss of time.
5. It is certain that these things will advance in price before long, so that we heartily advise you to buy as much as you can.
6. We shall be pleased to receive your order.
7. Owing to the rush of orders, the stock of these goods has been exhausted, and we would recommend an excellent substitute.
8. More inquiries are being received for this commodity, which we believe is an indication of an increasing demand in your market.
9. Owing to the increased demand for these models, our stocks have almost been exhausted.
10. We are sorry to learn that you find our quotation of model no. 123 too high.
11. As our stocks of these goods are limited, we suggest you place an order immediately.
12. As the prices quoted are exceptionally low and likely to rise, please place your order without delay.
13. As soon as our present stock has run out, we shall have to revise our prices.
14. Please confirm your order at the price quoted.
15. We trust that you will take advantage of this seasonal demand and favor us with an order.
16. We are sure that these goods will meet your requirements and hope to receive your order.
17. This item is in huge demand, and we would advise you to place an order at once.
18. If our proposal is acceptable, please confirm by return mail.
19. Please let us know of your interest and comments soon.

練習題

1. 將下列的句子翻成英文

A. 我今早剛從展覽回來，發現貴公司尚未回覆我不在時所發的數通電子郵件。

B. 有好一段時間過去了，我們將會感謝您的任何回音。

C. 參照我方的紀錄，我們已有一段時間未曾和您聯絡了，我們誠心希望貴公司生意興隆。

D. 我們於 9 月 1 日寄給貴公司我們商品最新的目錄與最好的價格，但是至今尚未收到貴公司的回音，不知貴公司是否完好收到以上。

E. 保持不斷通知新開發的商品給像貴公司這樣的好客戶，我們相信您應已收到我們 5 月 1 日寄去的新單頁目錄。

F. 請注意我們最近得到大量訂單，您最好能儘快確認訂單，以便我們可以為您預留生產期。

G. 由於這項產品的大量需求，我們建議您儘快下單。

H. 雖然價格從 2 月份以來逐步上升，我們仍盡力維持我們的報價。我們希望您能在成本不可避免的進一步上升之前把訂單下給我們。

I. 由於我們的生產排程很緊，我們會建議您在這個月月底之前給我們訂單。

J. 我們向您保證品質和我們寄去的樣品一致，您能儘快下單，我們將不勝感激。

K. 我們的新產品賣得非常好，我們很有信心向您推薦，我們最近已經收到很多訂單了，建議您儘快下單。

2. 將下列中文書信翻譯成英文

本公司已於 4 月 15 日寄給貴公司我們最新的女性流行鞋的型錄，不知貴公司是否已經收到。

這些都是今年最新流行的款式，非常受到消費者的喜愛，如果貴公司有興趣任何款式，請儘快下訂單。如有需要樣品，請來函告知，我們將樂意並儘快提供 2 雙免費的樣品供客戶評估。

希望儘速能得知貴公司的興趣與意見。

3. 寫一封追蹤信，表示我方已於 6/18 寄給客戶報價與樣品，但是一直到今日尚未收到回音，想請問有沒有收到樣品，看看客戶有無任何意見。如樣品不符客戶需求，我們也可根據客戶的規格打樣，再送客戶所需的規格樣品。

4. 寫一封對老客戶的追蹤信，表示距上次訂單已有一段時間，希望不是因為客戶不滿我方的服務。我們保證客戶在和我們交易後都會非常滿意，如果您未下單是因為處理的商品種類改變，請讓我們知道，看看是否可以滿足您的需求。希望您還是銷售同款運動商品，因此我們在隨函附上最新的型錄。我們認為這份型錄在產品系列、品質和價格上，比其他業者優異。您從目錄中可看到，隨著匯率管制以及官方措施的取消，與上次訂單相比，我們的交易條件優惠許多。

5. 寫一封參展回來後的追蹤信，表示在 CES 展覽後，感謝對方來訪問我們的攤位，問其看過目錄後的意見或是興趣，再次強調我們產品的特色，最後希望有機會建立商業關係。

CHAPTER

09 訂 單

Purchasing Order

9.1 買方下單

買賣雙方在價格談妥，樣品規格確認後，買方就會決定下訂單給賣方，並與賣方確認採購合約與付款及出貨的方式。買方正式下單的訂單名稱有幾種：

- Trial order/sample order（試銷訂單／樣品單）：這是一種少量、單一次的訂單。
- Formal order（正式訂單）：這是買方開始建立常態採購訂單的第一張訂單。
- Repeat order（追加訂單）：買方在第一次購買後，對曾經買過的商品再一次進行購買的訂單。
- Regular order（定期採購訂單）：買方確定與賣方的長期性、持續性與定量的購買訂單。一般的長期性訂單可分為季訂單、半年單或是一年單，一次下定大量的訂單再根據需求量要求賣方依據一定的時間逐批出貨。

買方對賣方的採購活動有二種情形：

1. 買方發出採購訂單 (purchase order, P/O)

買方在決定取得賣方商品時，有可能下購買樣品或是小量的訂單，此時買方會發出採購的訂單，此種購買有可能是一次性的購買，買方會用其制式的「訂購單」並於訂購單上載明產品明細、交期、付款條件、交易條件、數量、價格…等明細。並將正本訂單寄給賣方，請賣方簽名寄回副本。因為訂單屬於合約的一種，如要有法律的效用，一定要雙方簽名的正本才有效。

有時候買方在採購樣品單時，或是只有一次性購買項目少樣的商品時，可能不會下訂購單而是直接於 e-mail 上載明。待第一次購買的商品於市場上反應良好後，才會再用正式的訂購單購買多樣項目及數量大的訂單。這時賣方才正式成為買方的供應商 (supplier)。

2. 雙方簽定長期合約 (purchase/sales contract)

買方經過長期的購買後，如果其中的供應商的品質、價格、交期與困難排除的能力等經考核之後，認定可成為永續的供應商，則買方會與賣方簽訂長期合約，此時供應商即成為策略供應商 (strategic supplier)，此種契約的內容詳細且繁瑣，大多用於中、大型企業，交易性質複雜且金額較大的情況。有些大型企業與策略供應商的採購平台已不使用紙本「訂購單」，雙方會透過企業資源整合系統 (Enterprise Resource Planning, ERP) 於電腦上直接下訂單，而供應商會於此平台上回覆買方作訂單的確認，這樣的方式不但可以節省紙張的浪費，於訂單的下單與確認動作也非常迅速。

一般中小型企業多使用印好的制式「訂購單」，並且於訂購單的背面會列出訂單的條款 (terms and conditions)。訂單條款的內容可包含下列項目：

- 確定交易雙方身分
- 價格及付款方式
- 品質的要求、生產的排成與報告
- 檢驗的細節規定
- 包裝的規定
- 不良品的處理
- 專利與智慧財產權的規範
- 合約的期限
- 其他事項的約定

另外還有一些小型的公司並沒有印好的訂購單，而是以信函或是電子郵件上書寫訂單。在寄送這樣寫法的訂單時，務必包含下列項目，以確保訂購的正確和清楚：

- 產品編號
- 正確及完整的商品說明
- 數量
- 價格
- 交貨需求（地點、日期、運送方式等）
- 付款方式

範例 9-1　買方：於 e-mail 上直接下單

To: lindachen@tbi.com.tw
From: jackturner@pioneer.com
Cc:
Bcc:
Subject: Purchase order
Attachment:

Dear Linda,

Please supply the following items:

Item number	Description	Quantity	Unit Price
JIB225	Window cutting machine	8	US$2,020.00
CC623A1	Bench Lathe	1	US$ 750.00
			Total:$16,910.00

Payment terms shall be standard 2%-10/net 30.
Please ship the items no later than March 29, 2021 by using UPS expedited shipping.

Ship all items to: Receiving Office
Pioneer Corporation
3080 Bowers Avenue
Santa Clara, CA 95054, USA
Tel: (408) 576-8888

Best wishes,

Jack Turner
Purchasing Manager

整「2%-10/net 30」的意思是，如果客戶於 10 天內付款，就可以得到 2% 的現金折扣，如果 10 天內沒付款，最遲須於 30 天內付清全額。

翻譯

請提供以下商品：

商品編號	說明	數量	單位價格
JIB225	Window cutting machine	8	US$2,020.00
CC623A1	Bench Lathe	1	US$ 750.00
			Total:$16,910.00

付款條件為：10 日內付款有 2% 現金折扣，30 日內付清。
請於 2021 年 3 月 29 日前使用 UPS 運送貨物。

請將所有商品寄到：收貨辦公室
先鋒公司
3080 Bowers Avenue
Santa Clara，CA 95054，USA
電話：(408)576-8888

範例 9-2　買方：接受報價並下訂單

To	lindachen@tbi.com.tw
From	jackturner@pioneer.com
Cc	
Bcc	
Subject	Purchase order No. PO1265
Attachment	P/O1265.doc

Dear Linda,

Thank you for your quotation Q123. We accept your price and terms and would like to place order no. PO1265 as attached. The formal order sheet will be followed by post.

As agreed, please do 100% test before shipment to make sure the quality and ship the goods on time. Please confirm your acceptance by return e-mail and send us your Proforma Invoice for opening L/C.

Best wishes,

Jack Turner
Purchasing Manager

L/C 是指信用狀 (Letter of Credit) 為國際貿易中買方付款的常用方法之一。

翻譯

感謝您的報價單 Q123。我們同意接受您的價格和條款。附件是我們的訂單號碼 PO1265。正式的訂單將隨後寄給你。

按照約定，請在發貨前做 100% 的測試，以確保質量並準時交貨。請回函確認您的接受，並寄給我們預付發票寄以利開立信用狀。

附件：訂單 PO1265

Pioneer Corporation
3080 Bowers Avenue
Santa Clara, CA 95054, USA
Tel: (408) 576-8888

Date: March 10, 2021
PO No.1265

PURCHASE ORDER

To: Taiwan Bear International
3F, No. 11, Park Avenue II
Science-Based Industrial Park
Hsin-Chu 30075, Taiwan
Attn: Ms. Linda Chen
Your ref: Q123

Shipping Mark:

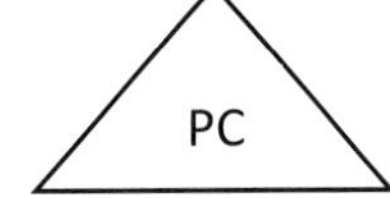

C/No.1-6
Made in Taiwan

Item No.	Specification	Qty	U/P	Total Amount
SR661	Actuator Ex-Proof, 3 Phase Motor 220V, 60Hz	6 sets	FOB Taiwan US$800.00	US$4,800.00

Total Amount: USD Four Thousand and Eight Hundred only.
Packing: By standard export wooden case.
Shipment: 45 days upon receipt orders
Payment: by confirmed and irrevocable L/C at sight in favor of Taiwan Bear International

Note: Following documents are required with shipment: (註1)

1) Warranty letter to certify the guarantee period of one year
2) Original manufacturer certificate
3) Material certificate
4) Inspection and testeport
5) Operation and maintenance manual (Instruction Manual)

The Buyer
Pioneer Corporation

Jack Turner

Jack Turner/Purchase Manager

The Seller
Taiwan Bear International

Linda Chen/Sales Manager

註 1：出貨文件通常會在新客戶（買方）下訂單給賣方後，根據買方國家的進口要求與客戶本身的需要，買方會將文件的細部準則與規定另外寫信告知賣方。如果已經合作很久的老客戶，對於文件的要求只在信用狀上說明即可。

範例 9-3　買方：樣品確定接受報價，並下訂單

To	lindachen@tbi.com.tw
From	jackturner@pioneer.com
Cc	
Bcc	
Subject	PO KK2017008
Attachment	PO-KK2017008.doc; instructions.doc

Dear Linda,

We have approved your samples and quotation for which we thank you and would like to confirm our order as per the attached list. The special instructions are listed in the enclosed order no. KK-2017008. Please confirm so that we can open the irrevocable L/C with you.

If this initial order turns out to be satisfactory, we shall be able to give you a large order in the near future. Thank you for your attention and we look forward to receiving your confirmation by return P/I.

Best wishes,

Jack Turner
Purchasing Manager

翻譯

我們已經審核通過您的樣品和報價，在此感謝您的報價與樣品，並附上我們的訂單。在訂單號碼 KK-2017008 中有訂單的特別說明。請確認訂單，以便我們開發不可撤銷的信用狀給您。

如果這個初始的訂單的結果令人滿意，我們將在不久的將來再次下一個大訂單。感謝您的關注，希望能儘快收到您的預付發票以確認訂單。

範例 9-4　買方：樣品確定接受報價，並下訂單

To lindachen@tbi.com.tw
From jackturner@pioneer.com
Cc
Bcc
Subject PO KK2017008
Attachment PO.doc; Shipping Instructions.doc

Dear Linda,

Thank you very much for your quotation of Feb. 11. We have studied your price and catalog and have chosen some items for which we enclose our order.

We would like to point out that this is a trial order and if we are satisfied with the quality of your goods and the service you render to us, you can expect regular repeat orders from us. To avoid any difficulties with our customs authorities, please make sure that our shipping instructions are carefully observed.

For our credit standing we refer you to Bank of America in LA. Please confirm the acceptance of our order.

Best wishes,

Jack Turner
Purchasing Manager

翻譯

非常感謝您 2 月 11 日的報價。我們已經研究了您的價格和目錄，並選擇了某些商品，附件是我們的訂單。

我們想跟您說這是一張試用訂單，如果我們對您的商品與提供的服務感到滿意，我們以後就會下經常性的訂單。為了避免我們雙方在海關當局遇到任何困難，請仔細研讀我們的出貨說明。

有關本公司的信用狀態，我們可以提供您洛杉磯的美國銀行作為信用調查。請確認接受我們的訂單。

9.2 賣方接受訂單

賣方在收到訂單後，應立刻回信通知買方，如果於訂單上有任何問題，也應儘速與買方釐清。

賣方對訂單的確認，需於正本訂單上簽名並寄回副本，這一份則與正本訂單的法律效用一樣。如果買方沒有寄來正本訂單，或者買方需申請輸入許可證時，賣方要寄「預付發票」(Proforma Invoice, P/I)，載明商品明細、交期、付款條件、數量、包裝等細節，請客戶簽名確認後，賣方再以 P/I 開出相關的生產單需求給生產部門。此份 P/I 經客戶簽名確認後，同樣與正本訂單一樣具有法律效用。

範例 9-5　賣方：回覆接受訂單（Re 範例 9-1）

To	jackturner@pioneer.com
From	lindachen@tbi.com.tw
Cc	
Bcc	
Subject	Re:Purchase order
Attachment	

Dear Jack,

We thank you for your reply of Feb. 14. Our company is very glad to receive your order for 8 Window Cutting Machines and one Bench Lathe. Special care will be devoted to the execution to your order. We will notify you as soon as the consignment is ready for transport.

We hope this first order will lead to further business.

Best regards,

Linda Chen
Sales Manager

翻譯

感謝您 2 月 14 日的回覆。我們公司很高興收到您的訂單，訂購 8 台窗戶切割機和一台車床的訂單。我們將特別關注並執行您的訂單。一旦貨物準備出貨，我們會儘快通知您。

我們希望這第一筆訂單將會使我們雙方的業務更進一步。

範例 9-6　賣方：回覆接受訂單並附上 P/I（Re 範例 9-2）

To	jackturner@pioneer.com
From	lindachen@tbi.com.tw
Cc	
Bcc	
Subject	Re:Purchase order No. PO1265
Attachment	P/I No. 7211

Dear Jack,

Many thanks for your e-mail of Mar. 10 and confirm the acceptance of your order no. 1265. We also confirm that we will do 100% test before shipment and will proceed as stipulated in the order.

The duly signed P.O. is delivered to you by separate mail. Enclosed please find our Proforma Invoice No. 7211 as requested for opening L/C.

We look forward to fostering and strengthening our future business cooperation.

Best regards,

Linda Chen
Sales Manager

翻譯

非常感謝您 3 月 10 日的來函，在此確認接受您的訂單號 1265。我們並向您確認我們將在發貨前進行 100% 的測試，並將按照訂單中的規定進行。

訂單我們已簽名確認並郵寄回去給您。隨函附上開立信用狀所要求的預付發票，號碼：7211。

我們期待雙方未來的業務合作。

附件：預付發票樣本 (Proforma Invoice)

Taiwan Bear International
3F, No. 11, Park Avenue II
Science-Based Industrial Park
Hsin-Chu 30075, Taiwan

Date: March 12, 2021
P/I No.7211

PROFORMA INVOICE

To: Pioneer Cooperation
3080 Bowers Avenue
Santa Clara, CA 95054, USA
Attn: Mr. Jack Turner

Your Ref: PO1265

Item No.	Specification	Qty	U/P	Total Amount
			FOB Taiwan	
SR661	Actuator Ex-Proof, 3 Phase Motor 220V, 60Hz	6 sets	US$800.00	US$4,800.00

Total Amount: USD FOUR THOUSAND AND EIGHT HUNDRED ONLY.
VVVVVVVVVVVVVVVVVVVVVVVVVVVVVVVVVVVVVVVVVVVVVVVVVV

The Seller
Taiwan Bear International

Linda Chen

Linda Chen/Sales Manager

The Buyer
Pioneer Corporation

Jack Turner/Purchase Manager

範例 9-7　賣方：回覆接受訂單，並告知調整包裝，且附上 P/I（Re 範例 9-3）

To: jackturner@pioneer.com
From: lindachen@tbi.com.tw
Cc:
Bcc:
Subject: Re: PO KK2017008
Attachment: P/I No. 7212

Dear Jack,

Thank you for your e-mail dated Sep. 21 confirming PO KK2017008 for a full 20' container. Attached please find our PI#7212. Meanwhile, please read the following explanations:

1. As PD-3883 is regularly packed in 8 sets/ctn, we included 2 additional sets to the original quantity. Therefore, there are 100 sets for this item. We hope you can understand and accept this arrangement.
2. According to your order quantity, the total measurement is about 860', so we will fill the rest of the container with 60 sets of PD-1688 to effect a full 20' container shipment.

If everything is satisfactory, please kindly confirm by signing our P/I return. We look forward to your prompt confirmation.

Best regards,

Linda Chen
Sales Manager

有時客戶下單，數量無法剛好符合裝箱，可能只差一些數量，此時賣方可主動幫客戶調整，並於訂單確認前告知客戶。另外，如果數量只差一些可達滿櫃或是客戶定的數量不足滿櫃卻要以滿櫃出貨，也應告訴客戶，請客戶增加訂購量。

翻譯

感謝您於 9 月 21 日來函，確認訂單號碼 KK2017008，並須為為 20 呎櫃的出貨。在此附上我們的預付發票，號碼 7212。同時，下列是我們的說明：

1. 由於 PD-3883 通常以 8 套／箱的形式包裝，因此我們在原始數量中增加了 2 套。這樣，這項商品將有 100 套。我們希望你能理解並接受這種調整。
2. 根據您的訂單數量，總才積約為 860 才，因此貨櫃剩下的才積我們將裝運 60 套的 PD-1688，以實現 20 呎櫃的整櫃出貨。

如果以上一切您都可以同意，請於我們的預付發票上簽名確認，並寄回。我們期待您的快速確認。

附件：預付發票樣本 (Proforma Invoice)

PROFORMA INVOICE

Messrs.: Taiwan Bear International
3F, No. 11, Park Avenue II
Science-Based Industrial Park
Hsin-Chu 30075, Taiwan

Date: Sep. 23, 2021
Customer PO#: KK2017008
Ref: 20171213

Payment: By Irrevocable L/C at sight in our favor
Shipment: Around Nov. 23, 2021
From Taiwan to L.A.
Terms: C&F L.A.

Item No.	Specification	Qty	U/P	Total Amount
PD-3883	Kitchen Knives 10 packs/CTN 1.8’	2040 packs	US$2.28	US$4,651.20
PD-1688	Blades for PD-3883 100 sets/CTN 1.3’	1500 sets	US$0.80	US$1,200.00
Total: 40.70 CUFT		2040 packs & 1500 sets vvvvvvvvvvvvvvvvvvvvv		US$5,851.20 vvvvvvvvvvvvv

SAY TOTAL U.S. DOLLARS FIVE THOUSAND EIGHT HUNDRED FIFTY ONE AND CENTS TWENTY ONLY.

Shipping mark:

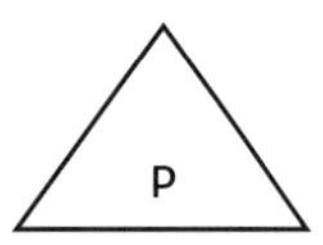

C/No.1-6
Made in Taiwan

The Seller
Taiwan Bear International

Linda Chen

Linda Chen/Sales Manager

The Buyer
Pioneer Corporation

Jack Turner/Purchase Manager

9.3 賣方不接受訂單

有時候賣方可能有一些情形而無法接受訂單，這些情形可能有下列幾種情況：

1. 該訂購的商品已售完（可推薦替代品）
2. 該產品已停產（可推薦新商品）
3. 客戶訂購量不足
4. 交貨期太趕
5. 付款方式無法接受
6. 前帳未清，欠款太多，無法再接受其訂單

賣方在寫回絕訂單的信函時，必須格外小心，以免影響雙方良好的關係以及未來的交易機會。

範例 9-8 賣方：無法接受訂單，因已不生產

To	jackturner@pioneer.com
From	lindachen@tbi.com.tw
Cc	
Bcc	
Subject	Re: PO KK2017008
Attachment	

Dear Jack,

Thank you for your e-mail dated of Jan. 18 in which you placed an order for our sweater model 551M. We appreciate your interest but regret to inform you that we no longer produce this item. We put forward to our new model 663M for your consideration, which we feel is an excellent replacement.

The quality and design are greatly improved at very little extra cost and are enjoying excellent sales. Concerning the price, we offer you the same price as the older model if you can place an order exceeding US$10,000. If your order does not exceed this amount, the price will be an additional 5%. We have confidence in this new model and are sure that it will give you complete more satisfaction.

We have sent you a sample by Federal Express today. We hope you will receive it before this Friday and will confirm an order soon.

Best regards,

Linda Chen
Sales Manager

翻譯

感謝您在 1 月 18 日發送的來函，並下單訂購我們的毛衣型號 551M。我們非常感謝您對我們商品的興趣。但是在此很遺憾地通知您，我們已不再生產這個產品。我們目前有新的新型號 663M，請您看看是否合用，我們認為這是一個很好的替代品。

此一款的品質和設計大大地改善，但是卻只增加很少的額外成本，並且市場的銷售量很好。關於價格，如果您可以下單超過 10,000 美元，我們可以維持與舊款相同的價格。如果您的訂單沒有超過這個數額，價格將是舊價還需增加額外的 5%。我們對這個新款式充滿信心，並確信它會讓您完全滿意。

我們今天已用聯邦快遞寄給您一個樣品。我們希望您能在本週五之前收到它，並很快確認訂單。

範例 9-9　賣方：無法接受訂單，因出價太低

To jackturner@pioneer.com
From lindachen@tbi.com.tw
Cc
Bcc
Subject Re: PO 2017112
Attachment

Dear Jack,

We are delighted to receive your PO No. 2017112 for 2000 dozens of kid's sweater. However, we could not accept this order at the price requested.

You could find in our quotation in which we offered you the best prices. Since then, prices have tended to rise rather than fall, and our profit margin does not allow us any concession by way of discount of prices. Therefore, we could not offer you the additional 5% discount.

We will be glad to execute your order if you can accept the price of US$22.35 per dozen FOB Taiwan. Please advise in your early convenience.

Best regards,

Linda Chen
Sales Manager

翻譯

我們很高興收到您的訂單，號碼 2017112 訂購 2000 打的小孩子毛衣。但是，我們無法按所要求的價格接受此訂單。

您可以在我們的報價中發現我們已提供最優惠價格的報價。在我們報價後，價格趨勢就往上漲而不是下跌。而且我們的毛利率也已不允許我們在給任何的折扣。因此，我們無法提供額外的 5% 折扣。

如果您能接受每打 FOB 台灣 22.35 美元的價格，我們將很樂意確認您的訂單。請儘早告知。

範例 9-10　賣方：無法接受訂單，因數量太少

To	jackturner@pioneer.com
From	lindachen@tbi.com.tw
Cc	
Bcc	
Subject	Re: PO 2017112
Attachment	

Dear Jack,

Thank you for your PO#2017112 in which you placed a further order for 50 units of PA-1566 as per the original sample order.

We regret to inform you that we cannot accept such a small order quantity. The original 100 units we made for you were intended as a service in the hope that the trial order would lead to more substantial orders. For this additional 50 units, we don't currently have any stock and have to produce on line again. The minimum production quantity is 500 units, therefore we are unable to accept this order.

Please kindly understand our position and we hope you will place an order for more than 500 units in the future.

Best regards,

Linda Chen
Sales Manager

翻譯

感謝您的訂單，號碼 2017112。您根據樣品訂單而來下此訂單，訂購 50 件 PA-1566。

我們很遺憾地通知您，我們不能接受這麼小數量的訂單。我們為您製作的最初的 100 件的樣品單是為了服務客戶，希望這一試用訂單之後能有更多的訂單。對於這個額外的 50 件，我們目前沒有任何庫存，必須再次安排生產。因為最小生產數量為 500 個單位，因此我們無法接受此訂單。

請您理解我們的困難，我們希望您將在未來的訂購量能超過 500 個單位。

範例 9-11　賣方：無法接受訂單，因無庫存

To	jackturner@pioneer.com
From	lindachen@tbi.com.tw
Cc	
Bcc	
Subject	Re: PO 2017112
Attachment	

Dear Jack,

Many thanks for your order dated March 20.

Unfortunately, we are out of stock of the Woman Sweater (item no. WS-445). This is due to the prolonged cold weather, which has increased demand considerably. The manufacturers have, however, promised us a further supply by the end of this month. If you can wait until then we will send you the goods immediately.

Please let us have your instructions.

Best regards,

Linda Chen
Sales Manager

翻譯

非常感謝您 3 月 20 日的訂單。

不幸的是，我們目前庫存上沒有女款毛衣（商品編號 WS-445）。由於長時間的寒冷天氣，導致需求量的大增。我們的製造商已經承諾在本月底前可以提供商品。如果您可以等到那時候，那麼我們可以在那時立即將貨物寄給您。

請告知您的決定。

範例 9-12　賣方：無法接受訂單，因無法接受付款條件

To jackturner@pioneer.com
From lindachen@tbi.com.tw
Cc
Bcc
Subject Payment terms for PO 2017112
Attachment

Dear Jack,

We are glad to learn from your e-mail of May 5 that you are going to place an order for our product quoted on April 11.

However, regarding the terms of payment, we would like to inform you that we cannot accept the D/P terms because it is contrary to our company policy. We only accept an irrevocable L/C or T/T in advance before shipment. If it is inconvenient to open a L/C in your country, we suggest you remit by T/T. As to the sample order, we will accept your bank draft. We feel that this method of payment (T/T for regular orders and bank draft for sample orders) will be the most convenient for both of us.

Your kind attention and acceptance with respect to the above will be appreciated. We look forward to your return confirmation soon.

Best regards,

Linda Chen
Sales Manager

翻譯

我們很高興從 5 月 5 日的來函中得知，您將根據我們 4 月 11 日寄給您的報價單下一張產品訂單。

但是，關於付款條件，我們想通知您，我們不能接受付款交單的條件，因為這違反了我們公司的政策。我們接受不可撤銷的信用狀或是出貨前電匯。如果在貴國開立信用狀不方便，我們建議您以電匯方式付款。至於樣本訂單，我們可以接受銀行匯票。我們認為這種付款方式（正常訂單的用電匯和樣本訂單用銀行匯票）對我們雙方來說都是最方便的。

感謝您對上述內容的關注和接受。我們期待能收到您的回覆確認。

範例 9-13　賣方：無法接受訂單，因積欠貨款未清

To: jackturner@pioneer.com
From: lindachen@tbi.com.tw
Cc:
Bcc:
Subject: Re:PO#2017112
Attachment:

Dear Jack,

We were pleased to receive your order PO#2017112 for further supply of Coffee Grinders and Bread Makers.

At present the balance of your account stands at US$3,182.54, and unfortunately we are unable to grant credit for further supplies until this balance is reduced. I hope you can appreciate that we do have commitments of our own to meet, and we can only do this by asking our customers to keep their accounts within reasonable limits.

If you can make payment for at least US$1,500.00 before this Friday, we will be happy to supply these new goods immediately.

Thank you for your understanding.

Best regards,

Linda Chen
Sales Manager

翻譯

我們很高興收到訂單，號碼 2017112，訂購咖啡研磨機和麵包機。

目前您的賬戶餘額為 3,182.54 美元，很不幸的，我們無法再出貨給您，除非您能先結清貨款。我希望您能夠理解我們對做事業的承諾。因此，我們也希望客戶也能維持承諾，所以我們客戶的帳戶餘額將維持在合理的額度內。

如果您可以在本週五之前支付至少 1,500 美元，我們將很樂意立即提供您所需的商品。

感謝您的理解。

常用句

買方

1. Please supply the goods listed below as per your quotation and samples submitted on October 20.
2. Please kindly acknowledge this order and confirm that you will be able to delivery by the end of this month.
3. We have placed a trial order. If the quality is up to our expectations, we will make further orders in the near future.
4. Enclosed please find PO No. 221 for cutting machine.
5. We thank you for your quotation of March 3, 2021, and should be glad if you would accept our order for the following items.
6. We have the pleasure of placing with you a formal order for these goods.
7. We trust that this trial order will lead to further dealings between our two companies.
8. Your prompt attention to this order will be appreciated.
9. As the goods are urgently required, we should be grateful for delivery by June 10.

賣方

1. We shall be pleased if we receive your initial order soon.
2. We thank you for your order no. PO121.
3. We are pleased to accept your order and attached our confirmation of P/I No. PI456.
4. We hope to have your order before further rises in prices.
5. Any order from you will receive our special attention.
6. If you want to deliver goods by end of Sep., you have to place orders before end of July.
7. Our stock is short because of the high demand this season.
8. Orders are rushing in and our stocks are running out quickly.
9. The goods you inquired are sold out, but we can offer you a substitute.
10. The goods you inquired are no longer manufactured, and we recommend the excellent substitute which is item no. PA-15.
11. We can offer you a substitute which is the same price and of similar quality to the goods ordered.
12. We regret to inform you that the goods ordered are out of stock.
13. The price does not induce us to manufacture the goods.
14. Our factory is fully employed, and our output capacity cannot keep pace with demand. So, we would prefer not to take any more orders for the time being.
15. We are unable to promise shipment within one month.

練習題

1. 將下列的句子翻成英文

A. 因市場需求增加，我們要追加 6,000 個，請修正訂單數量並確認。

B. 很抱歉，因為景氣不好，我們想減少數量 1,000 個。

C. 我們同意接受，請修正訂單數量。

D. 很抱歉無法接受增加數量，因為生產已經完成，所增加數量太少，很難備料及生產。

E. 請就報價確認訂單。

F. 貴公司若認為條件可以接受，即請確認有關上述修正的訂單。

G. 貴公司下給我們的任何訂單，我們都會迅速仔細處理。

H. 謝謝您的報價單 QQ123，我們接受貴方的價格及條件，並確認我們的正式訂單 no. AAA，影本如附，正本將郵寄過去。

2. 將下列中文書信翻譯成英文

我們接受貴公司報價單上的價格及交期，並很高興附上我們的新訂單號碼 AA-2017888. 請務必準時出貨，因為此一訂單是我們要趕在耶誕節銷售的商品。

請確認定單，並傳預付發票給我們，以便開立信用狀。

3. 賣方：請就以上的信函，回一封完整的確認函。

4. 請寫一封拒絕接收訂單的信函，表示對客戶的訂單的歡迎，但是所定的商品本公司已不生產。我們有新的商品，品質更好，功能更佳，雖然價格會高一些，但是可以推薦給客戶作為替代品的考量。

5. 請寫一封拒絕接收訂單的信函，客戶下訂單但是付款方式為出貨後 60 天電匯，說明我們的付款條件為出貨前全額付清或是信用狀方式。

CHAPTER

10 徵信

Credit Inquiry

範例 10-1　賣方：請客戶提供企業備詢人

範例 10-2　賣方：寫信給企業備詢作客戶徵信 I

範例 10-3　賣方：寫信給企業備詢作客戶徵信 II

範例 10-4　賣方：請客戶的往來銀行協助徵信

範例 10-5　賣方：委託商業徵信所進行徵信

範例 10-6　賣方：寫給新客戶進行徵信

信用調查的信件是用來收集客戶的信貸資料，藉以了解客戶的信用狀態以及是否有償債的能力。當新的客戶下單時，供應商會要求對方提供一些有關信用評等的資訊，尤其是客戶下了大筆的訂單時。

信用調查的信基本上有兩種：一是寫給客戶所提供的信用擔保人，一是寫給客戶直接詢問。

客戶提供的備詢來源包含：

- 客戶提供的備詢商業往來企業 (trade reference)
- 客戶的往來銀行 (trace bank)
- 各種商會 (trade association)
- 商業徵信所 (credit inquiry agency)

由於這些參考的備詢企業都是客戶提供，因此在採用時必須謹慎小心，畢竟只有說話對客戶有利的公司才會被列為備詢企業。即使客戶提供的往來銀行也有可能誤導，客戶可能擁有一個相當令人滿意的銀行帳戶，實際業務卻大有問題的情況也有可能發生。

寫信給備詢人時，措詞必須正式有禮，通常需包含下列幾點：

- 提供客戶的背景
- 請備詢人就客戶的信用狀況提出看法
- 向備詢人保證會對其所提供的資料嚴加保密
- 附上回郵信封

一般說來這類徵信信函的收件人多是企業的高階主管，因此信上需標示「機密」(Confidential) 字樣，當我們收到對方回覆的信函時，應該回信確認並且致謝，以示禮貌。

寄給客戶本身的信用調查信件就需要比較有技巧的寫法，尤其是要得體周到，要讓客戶覺得受到重視，而非質疑對方的信用狀態。信件的內容應該說明下列幾點：

- 謝謝客戶的訂購
- 說明信用調查是公司的例行步驟，而非不信任的象徵
- 說明公司信用徵信的政策與執行程序
- 讓客戶了解，儘速提供信用資料對客戶本身有益，有助於公司迅速處理客戶的訂單

範例 10-1　賣方：請客戶提供企業備詢人

To	jackturner@pioneer.com
From	lindachen@tbi.com.tw
Cc	
Bcc	
Subject	Trade reference
Attachment	

Dear Jack,

We were pleased to receive your first order with us dated Oct. 17. When opening new accounts, it is our usual practice to ask customers for trade reference. Please give us the names and addresses of two other suppliers or customers with whom you have regular dealings.

We hope to hear from you soon. Meanwhile, your order has been put in hand for production immediately after we hear from you.

Best regards,

Linda Chen
Sales Manager

翻譯

我們很高興在 10 月 17 日收到您的第一張訂單。對於新客戶開設新賬戶時，我們通常會要求客戶提供信用備詢的備詢對象資料。請提供兩個貴公司的供應商或是有定期交易的客戶名單與地址以供備詢。

希望儘快收到您的回信。同時，我們也會著手準備您的商品，一收到您的回信就會立刻安排生產。

範例 10-2　賣方：寫信給企業備詢作客戶徵信 I

Taiwan Bear International

3F, No. 11, Park Avenue II
Science-Based Industrial Park
Hsin-Chu 30075, Taiwan
Tel: 886-03-5798888
Fax: 886-035978891

February 19, 2021

Les Broome
Administrative Director
Exotic Textiles, Inc.
2060 Wasach Avenue
Olympia, WA 98501
USA

Dear Mr. Broome:

Pioneer Corporation wishes to open an account with us and have given your name as reference.

Please let us have your view on the firm's general standing and your opinion on whether they will be able to settle their accounts with a credit up to US$10,000 monthly.

We will, of course, treat all information in strict confidence. We enclosed a stamped, addressed envelope for your reply.

Sincerely yours,

Linda Chen

Linda Chen
Sales Manager

Enc.

範例 10-3　賣方：寫信給企業備詢作客戶徵信 II

Taiwan Bear International

3F, No. 11, Park Avenue II
Science-Based Industrial Park
Hsin-Chu 30075, Taiwan
Tel: 886-03-5798888
Fax: 886-035978891

March 15, 2021

Simon Hoover
General Manager
Long Wood Trade Co. Ltd.
366 Cherry St. London
England

Dear Mr. Hoover:

A purchase order from Pioneer Corporation in USA, for $5,000 worth of merchandise listed you as a credit reference.

We would appreciate any information you can provide on the credit history of Pioneer Corporation with your company. Key information would include how long the company has had an account with you and whether or not Pioneer Corporation has any outstanding debts. Be assured, we will keep any information you send us confidential.

Thank you for your assistance in this matter. Enclosed please find an addressed, postage paid envelope for your convenience.

Sincerely yours,

Linda Chen

Linda Chen
Sales Manager

Enc.

翻譯

先鋒公司希望我們開設一個業務記帳帳戶，並且給我們貴公司的名字作為信用備詢人。

請告知貴公司對先鋒公司經營狀況的看法，對於先鋒公司能否以每月約 10,000 美元的信用額結算業務帳戶給本公司一個參考的意見。

當然，我們會嚴格保密您所提供的所有信息。我們隨函附上了一封已付郵資並蓋有地址的回郵信封，供您回覆。

翻譯

美國先鋒公司向我們公司下了一張價值 5,000 美元的訂單，並將您列為信用備詢人。

我們希望您能提供關於先鋒公司與貴公司的信用往來的信息。這些信息包括先鋒公司與貴公司業務往來有多久了以及先鋒公司是否有任何未清償的債務。請放心，我們會將您給我們的任何信息都保密。

感謝您對此信用諮詢的幫助。隨函附上一封郵資已付的回郵信封，以方便您寄回函給我們。

範例 10-4　賣方：請客戶的往來銀行協助徵信

To ▷ servicer@firstbank.com

From ▷ lindachen@tbi.com.tw

Cc ▷

Bcc ▷

Subject ▷ Credit Inquiry

Attachment ▷

Dear Sir/Madam,

The Pioneer Corporation in US has asked for us to grant them a standing credit of $20,000. However, as our knowledge of this company is limited to a few months' trading on the basis of cash-on-invoice, we would like some information about their financial standing before proceeding.

The only reference they give us is that of their bankers – the Bank of America, L.A. We hope you can let us have any information about this company that would help us in our decision.

Best regards,

Linda Chen
Sales Manager

翻譯

美國先鋒公司要求我們給予他們 20,000 美元的信用額度以做為業務的往來使用。但是我們對該公司的了解僅限於幾個月的依發票付現的交易，因此我們希望先對對方的財務狀況有一些了解，再做後續處理。

他們給我們的唯一參考資料就是他們的往來銀行，位於加州的美國銀行。希望您能提供這家公司的任何信息，以幫助我們做出決定。

範例 10-5　賣方：委託商業徵信所進行徵信

To	servicer@creditagency.com
From	lindachen@tbi.com.tw
Cc	
Bcc	
Subject	Credit Inquiry
Attachment	

Dear Sir/Madam,

We have received a first order worth $10,000 from Pioneer Corporation, who has requested open account terms.

We have no information about this company, but as there are prospects of further large orders we should like to meet this order and provide open account terms if it is safe to do so.

Please let us have a report on the reputation and financial standing of the company and whether it is advisable for us to grant credit for this first order. We would also appreciate advice on the maximum amount for which it would be safe to grant credit on a quarterly account.

Thanks for your help in advance.

Best regards,

Linda Chen
Sales Manager

翻譯

我們收到了先鋒公司的第一筆訂單，價值 1 萬美元，先鋒公司要求以記帳方式供貨。

我們沒有關於該公司的任何資料，但是預期未來該公司會有更大的訂單，我們願意接受此訂單並同意記帳供貨的條款，想諮詢您一下看這樣的做法是否安全。

請您針對該公司的信譽和財務提供我們一份報告，也請您對於我們是否可以給第一筆訂單就接受記帳給予建議。另外，該公司如果是以季來結算業務，最多可以給多少信用額度才算安全，如果您能提供意見，我們將非常感激。

在此先感謝您的協助。

範例 10-6　賣方：寫給新客戶進行徵信

To jackturner@pioneer.com

From lindachen@tbi.com.tw

Cc

Bcc

Subject Credit Inquiry

Attachment

Dear Jack,

Thank you for your interest in Taiwan Bear International. Your order has received our prompt attention and we are eager to expedite this shipment as early as possible. We never forget that new customers like you are responsible for our continued growth.

With first time customers, our policy is to request the usual certified financial statements, references, and the name of your bank, in order to open your account. All the information you provide to us will, of course, be kept confidential.

We look forward to receiving your information as soon as possible so that we can process your account and order immediately. We will do everything we can to ensure that this is a long and mutually profitable relationship. Please don't hesitate to tell us how we can be of service to you.

Sincerely,

Linda Chen
Sales Manager

翻譯

感謝您訂購本公司的產品。我們已經開始處理您的訂單，希望能儘快出貨給您。我們一直都了解正是像您這樣的客戶讓我們持續成長。

對於新客戶，我們的例行步驟是要請客戶提供審定的財務報表、信用擔保人和銀行名稱，以便開立您的貿易帳戶。您提供的所有資料我們都會嚴格保密。

我們期待儘快收到您的資料，以便我們能夠儘速處理您的訂單和設置貿易帳戶。我們將盡全力使這次的交易成為長期而互利的合作關係。有任何我們可以提供協助的地方，請儘管告訴我們。

常用句

1. We hope you will supply the usual trade references so that we can consider open account terms.
2. We will be in touch with you as soon as references are received.
3. It is our usual practice to request references from new customers, and we hope to receive these soon.
4. ABC company has given us your name in connection with his application for open account terms.
5. We have received a large order from ABC, who has given your name as a trade reference.
6. We will appreciate any information you can provide.
7. Any information provided will be treated in strictest confidence.
8. It is our policy to request the following information from new customers.
9. Providing the following information will help us create your account.

練習題

1. 將下列的句子翻成英文

A. 對於新客戶，本公司的政策需要索取客戶的審定財務報表、信用擔保人跟銀行名稱。

B. 根據本公司的政策，我們要請您提供信用擔保人。

C. 客戶下大額訂單時，本公司的例行做法是請客戶提供最新的財務報表。

D. ABC 公司將您列為他們的信用擔保者。

E. 我們尊重貴公司的隱私，絕不會將您提供的資料洩漏給第三者。

F. 財務資訊將嚴格保密。

G. 若您能鼎力提供資料給我們，將感激不盡。

H. 您可以確信，我們自您那裡得到的任何資料，您絕對免責。

I. 您是否可以提供一些有關他們的商務活動及財務狀況的資料。

2. 將下列中文書信翻譯成英文

Pioneer Corporation 公司向我們下了一張訂單，該公司將您列為信用的諮詢人。

想請您提供 Pioneer Corporation 與貴公司交易之信貸歷史，包括他們公司在貴公司的貿易帳戶往來有多長的時間、是否有未付清債務等。您提供的資訊我們將嚴格保密。

感謝您的協助，隨信附上寫好地址的回郵信封。

3. 請寫一封向客戶往來銀行徵信的信函。

4. 請寫一封向商業徵信所對新客戶的信用徵信的信函。

5. 請寫一封向客戶所提供的備查人進行客戶信用徵信的信函。

CHAPTER

11 付款條件

Payment Terms

11.1 付款條件 (Payment Terms)

國際交易，常用有下列幾種付款方式：

1. 信用狀 (letter of credit，L/C)

信用狀是國際貿易中最常用的一種付款方，因為它能夠為買方和賣方都提供高度的保障。開立信用狀是買方的責任，為此項服務買方還要付給銀行一筆手續費。買方請銀行開立信用狀，信用狀的內容基本上就是保證在賣方開出發貨的證明文件後，一定會付款。接著買方的銀行把信用狀寄給賣方的銀行。信用狀裡列出發貨限制、送貨方式及其他交易資訊。賣方的銀行收到信用狀後，賣方就開始發貨（11.3 有信用狀內容詳細說明）。

2. 跟單託收 (documentary collections)

跟單託收的付款方式為買方和賣方就價格、保險、訂單和訂單其他內容簽訂合約。賣方把發票、保險證明、提貨單等文件收齊。把貨物交給運輸公司時，運輸公司會在提貨單上簽名。然後賣方將這些所有文件交給他的銀行。賣方的銀行接著會把這些文件和賣方開出的匯票轉給買方的銀行。這時視雙方談成的交易方式，買方要不就在看到這些文件時付款給他的銀行，稱為「付款交單」(documents against payment, D/P)，要不就簽署匯票，同意於一定期限內付款，稱為「承兌交單」(documents against acceptance, D/A)。一旦買方付款了，銀行把這些文件交給買方，買方就成為貨物擁有人。此時銀行同時保護雙方利益，而雙方都有風險，因為買方可以拒絕付款，而賣方可以發出品質不佳的商品。

3. 貿易帳戶／記帳 (open account, O/A)

貿易帳戶的付款方式對賣方來說風險最大，對買方來說風險最小。通常是雙方已經有穩定與長期的合作關係，或是母子公司或是集團內的公司，因為賣方信任買方付款的能力與意願，才會採取此種付款方式。此時買方的帳先記在帳戶上，等收到貨物後，雙方約定一定的期間買方才償付貨款。

4. 分期付款 (installment)

此付款方式多用於機器設備、模具、工程案或是土地、廠房買賣等，此付款方式可使買賣雙方互相牽制，買方就不用擔心貨到無法使用，賣方也不用擔心買方收貨不付款，雙方共同約定付款的次數與百分比。

5. 寄售 (consignment)

寄售是賣掉商品才付款的方式，沒賣掉就不付款。所以此種付款方式多用於大廠（資本雄厚）或是零售商。

6. 交貨時收取現金 (cash on delivery, COD)

此為一般國內交易的付款方式，即貨送到指定倉庫就收取現金貨款。

7. 下單時即支付現金 (cash with order, CWO)

此一方式多用於市場缺貨或是搶貨的狀態，買方只好拿現金去排隊。

11.2 付款方法 (Payment Methods)

除了信用狀以外，其他付款條件 (D/P, D/A, O/A, installment, consignment, COD, CWD) 的付款方式皆可採下列任一方法付款：

1. 電匯：T/T (Telegram Transfer, Wire Transfer)
2. 信匯：M/T (Mailed Transfer)
3. 銀行本票：Bank Check
4. 銀行匯票：Bank Draft
5. 郵政匯票：Money Order

電匯 (T/T) 是一種付款的方法而非付款條件，但是現在的國際貿易中很多人使用上直接說「付款方式用 T/T」，是指付款條件可能為出貨前要收到貨款，則電匯就是出貨前的電匯 (T/T in advance)；若是出貨後 30 天內要收到貨款，則電匯就是出貨後的電匯 (T/T within 30 days after shipment)。出貨前的電匯對賣方的風險最小，對買方的風險最大，因為買方需於出貨前就交付貨款。通常是雙方第一次交易或是小額交易時才採用此種付款方式；或是長期經營的供應商，雙方的信任度已建立，對於常態性購買訂單買方才會同意採用電匯方式。

付款方式	狀況及注意事項
出貨前 T/T	出貨前收到貨款才出貨，一切 ok。
出貨時 T/T	待出貨後一週之內會收到貨款，文件寄出，一切 ok。
出貨後 T/T	須於貨款預定到期日時，適時跟催。
出貨後 D/P	須注意客戶是否如期付款，適時跟催。
出貨後 D/A	須注意客戶是否承兌，到其時是否付款，適時跟催。
出貨後 O/A	須注意買方是否付款，須適時跟催。
出貨後 L/C	進行押匯後，一段時日後若開狀銀行挑瑕疵或拒付，須與客戶儘速溝通，請客戶趕快付款，贖單。

11.3 信用狀 (Letter of Credit, L/C)

買賣雙方簽定訂單後，雙方若約定用「信用狀」為付款條件，按「國際條約」規定：「買方須於簽訂買賣合約後 15 日內開出信用狀，或安排付款。」因此，對於新客戶賣方應注意去追蹤或催促，以避免造成損失。

一般國際貿易常用的付款方式除了信用狀之外，還有電匯 (T/T, wire transfer) 的方式。因為信用狀的銀行費用較高，所以多使用於訂單價值較高或是新客戶的付款方式。但是對於老客戶可能就會採用 T/T 的方式以節省銀行費用。

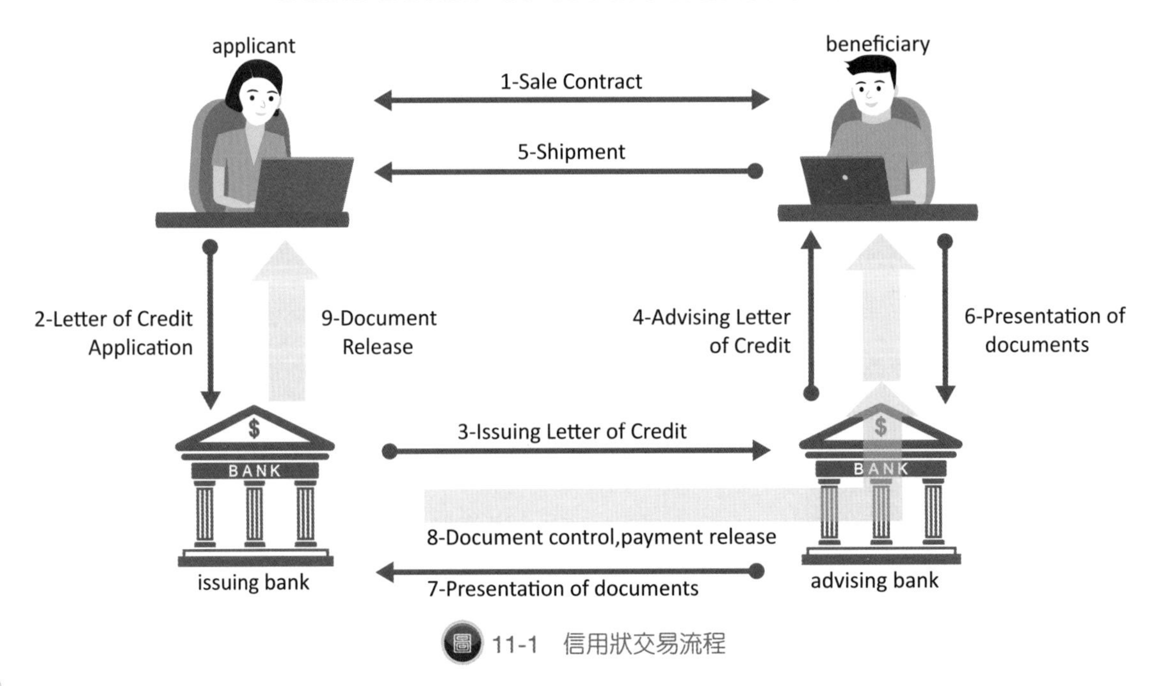

圖 11-1 信用狀交易流程

11.3.1 信用狀的內容

信用狀上的基本內容含下列項目：

1. 申請人 (Applicant/Drawee/Consignee)：買方公司名稱／地址。
2. 受益人 (Beneficiary/Drawer)：賣方公司名稱／地址。
3. 開狀銀行 (Opening/Issuing Bank)：買方往來銀行。
4. 通知銀行 (Advising/Notifying Bank)：通知信用狀的銀行，通常為開狀銀行的往來銀行。
5. 押匯銀行(Negotiating Bank):賣方的往來銀行,有時通知銀行與押匯銀行為同一家。
6. 保兌銀行 (Confirming Bank)：應開狀銀行的委託，就其所開的 L/C 承擔保證兌付的銀行，有時通知銀行與保兌銀行為同一家。
7. 補償／清算銀行 (Reimbursing Bank)：受開狀銀行的委託，對押匯銀行償還其押匯票款的銀行。
8. 付款銀行 (Paying Bank/ Drawee Bank)：大多為開狀銀行。
9. 轉交銀行 (Processing Bank)：對於有特別限定押匯銀行的信用狀，若受益人與此銀行並無往來，則須先向受益人有往來的銀行辦理押匯，再由該往來銀行將受益人之押匯資料轉交給銀行。再信用狀所指定的押匯銀行，則稱為「再押匯銀行」(Re-negotiating Bank) 或第二次押匯銀行。
10. 開狀日期 (Issuing Date)。
11. 信用狀號碼 (L/C No.; Credit No.)。
12. 金額 (Amount)。
13. 最後出貨日 (Latest Shipping Date)。
14. 有效期限 (Expiry Date for Negotiation)。
15. 提示押匯時間 (Drafts presented time for negotiation)。
16. 起運港 (Shipment from)。
17. 目的地 (To)。
18. 價格條件 (Price Terms)：有很多種像 Ex-work, FOB, C&F 或是 CIF。
19. 分批裝運 (Partial shipment)：同意 (Allowed) 或是不同意 (Prohibited)。

20. 轉運 (Transhipment)：同意 (Allowed) 或是不同意 (Prohibited)。

21. 貨櫃裝運 (Container Shipment)：必須 (Required)；准許 (Allowed)；不准 (Prohibited)。

22. 貨物內容 (Covering)。

23. 應檢附之單據 (Accompanied by the following documents)。

    必要單據有三項：

    - 商業發票 (Commercial Invoice)
    - 包裝單 (Packing List)
    - 提單：海運提單 (Bill of Lading, B/L)；空運提單 (Air Way Bill)

    另有其他文件依客戶要求再提示：

    - 保險單 (Insurance Policy or Certificate)：CIF 出貨才需要
    - 產地證明 (Certificate of Origin)：由商會發出
    - 受益人證明 (Beneficiary's Certificate)
    - 海關發票 (Special Custom Invoice, SCI)
    - 品質證明書 (Quality Certificate)
    - 其他還有檢驗報告、操作手冊、出貨通知等文件，視客戶要求照做即可

11.3.2 信用狀的種類

1. 依可否撤銷

(1) 可撤銷信用狀 (Revocable L/C)：在受益人未押匯前，買方可隨時片面修改或是撤銷，不須相關人等的同意。

(2) 不可撤銷信用狀 (Irrevocable L/C)：受益人收到信用狀後，非經受益人開狀銀行同意，不得隨意修改或撤銷。

2. 依有無保兌

(1) 保兌信用狀 (Confirmed L/C)：除開狀銀行外，另經一信譽良好的銀行保證付款得信用狀。

(2) 不保兌信用狀 (Unconfirmed L/C)：未經其他銀行附加保兌承諾的信用狀。

3. 依匯票期限

(1) 即期信用狀 (Sight L/C)：即受益人憑即期匯票押匯取款的信用狀。

(2) 遠期信用狀 (Usance L/C)：即受益人憑遠期匯票押匯取款的信用狀。

4. 依有無跟單

(1) 跟單信用狀 (Documentary L/C)：押匯時，須檢附單據。

(2) 無跟單信用狀 (Clean L/C)：押匯時，不須檢附單據。

5. 依可否轉讓

(1) 可轉讓信用狀 (Transferable L/C)：可轉給他人使用的信用狀，一般以一次為限。

(2) 不可轉讓信用狀 (Non-transferable L/C)：不可轉給他人使用的信用狀。

6. 依有無限定兌付銀行

(1) 一般信用狀 (General L/C)：即押匯時不限定押匯銀行，出口商多在自己的往來銀行押匯。

(2) 限押信用狀 (Restricted L/C)：即限定押匯銀行。

(3) 直接信用狀 (Straight L/C)：規定受益人只能向信用狀所指的銀行提示匯票或單據請求付款。

7. 其他種類的信用狀

(1) 國內信用狀 (Local L/C)：即開狀申請人、受益人與開狀銀行都在國內之信用狀。

(2) 對開信用狀 (Back-to-back L/C)：出口商依主信用狀 (Master L/C) 另開給工廠或供應商的信用狀，又稱 Secondary L/C。

(3) 循環信用狀 (Revolving L/C；Renewable L/C)：受益人在一定期間內及一定金額內可以循環反覆使用的信用狀。

(4) 擔保信用狀 (Stand by L/C)：專供保證用的信用狀。

(5) 紅條款信用狀 (Red Clause L/C)：受益人在未押匯前，即可預支款項。

11.3.3 信用狀的傳遞方式

信用狀的通知方式有三種方法：

1. 郵寄信用狀 (Mailed L/C)：即是正本信用狀。
2. 電報式信用狀
   (1) 簡電 (Short Cable, Brief Cable)：開狀銀行先將信用狀之主要內容電告通知銀行轉通知受益人（出口商），隨後正本再寄出。
   (2) 詳電 (Full Cable)：開狀銀行先將信用狀之全部詳細內容電告通知銀行，經通知銀行核驗押密碼後，轉通知受益人（出口商）。通常電文的前頁或是最後一頁都會記載「本電報為有效信用狀」(This cable is the operative instruction; An authenticated operative credit instrument.)
3. SWIFT 信用狀：為環球銀行財務電信協會 (Society for Worldwide Interbank Financial Telecommunication) 組織下的電訊系統，以現代傳訊方式所傳遞的信用狀，目前已逐漸取代傳統的郵寄與電報信用狀。

信用狀範例 1：

<BANK LETTERHEAD>

Date: **<DATE>**

Issued By: **<NAME AND COMPLETE ADDRESS OF BANK>**

Beneficiary: Texas Department of Licensing and Regulation
P.O. Box 12157
Austin, TX 78711

Applicant: **<NAME AND COMPLETE ADDRESS OF SERVICE CONTACT PROVIDER APPLICANT>**

Irrevocable Letter of Credit Number: **<NUMBER>**

Amount: **<$ AMOUNT>**

Effective Date: **< DATE>**

Expiration Date: **<DATE>**, or any automatically extended period thereafter.

We hereby issue our irrevocable letter of credit number **<NUMBER>** in your favor for the account of **<NAME OF SERVICE CONTACT PROVIDER APPLICANT>** for a sum not to exceed **<$ AMOUNT>**.

This irrevocable letter of credit is given as security for the benefit of a party who may suffer damages resulting from the failure of **<NAME OF SERVICE CONTACT PROVIDER APPLICANT>** to meet or perform its obligations under Texas Occupations Code, Chapter 1304, or the rules or regulations pertaining thereto.

We will honor your draft at sight drawn on **<NAME OF BANK>** in an amount not to exceed **<$ AMOUNT>** in the aggregate. Drafts hereunder must be marked "Drawn Under Irrevocable Letter of Credit No. **<NUMBER>**", and must be accompanied by a written statement from you stating that **<NAME OF SERVICE CONTACT PROVIDER APPLICANT>** failed to meet or perform its obligations under Texas Occupations Code, Chapter 1304, or the rules or regulations pertaining thereto. Such statement is to be signed by an authorized official of the Beneficiary.

It is a condition of this irrevocable letter of credit that it shall be automatically extended without amendment for an additional period of one (1) year from the original expiration date and each future expiration date, unless at least sixty (60) days prior to the then current expiration date, we send notice in writing to you that we elect not to automatically extend this irrevocable letter of credit for an additional one (1) year period. Notification will be sent to the Beneficiary at the address above and to the attention of "Compliance Division--Service Contract Providers Program."

We agree that we shall have no duty or right to inquire as to the basis upon which Beneficiary has determined to present to us any draft under this irrevocable letter of credit. Any draft(s) drawn under and in compliance with the terms and conditions of this irrevocable letter of credit will be duly honored. Multiple and partial drafts are permitted, not to exceed the aggregate amount of this irrevocable letter of credit.

Drafts on the irrevocable letter of credit shall be submitted to: **<NAME, ADDRESS AND PHONE NUMBER OF CONTACT PERSON AT THE BANK.>**

______________________________	______________________________
Signature of Authorized Bank Official	Title of Authorized Bank Official
______________________________	______________________________
Printed Name of Authorized Bank Official	Date

TDLR Form SCP 005 (6/2010)   This document is available on the TDLR website at www.license.state.tx.us.

信用狀範例 2：

INTERNATIONAL BANKING GROUP

ORIGINAL

Megabank Corporation

P.O. BOX 1000m ATLANTA, GEORGIA 30302-1000
CABLE ADDRESS: MegaB
TELEX NO. 1234567
SWIFT NO. MBBABC 72

OUR ADVICE NUMBER: EA00000091
ADVICE DATE:08MAR97
ISSUE BANK REF: 3312/ HBI/ 22341
EXPIRY DATE: 23JUN97

****AMOUNT****
USD****25,000.00

BENEFICIARY:
THE WALTON SUPPLY CO.
2356 SOUTH N.W. STREET
ATLANTA, GEORGIA 30345

APPLICANT:
HHB HONG KONG
34 INDUSTRIAL DRIVE
CENTRAL, HONG KONG

WE HAVE BEEN REQUESTED TO ADVISE TO YOU THE FOLLOWING LETTER OF CREDIT AS ISSUED BY:
THIRD HONG KONG BANK
1 CENTRAL TOWER
HONG KONG

PLEASE BE GUIDED BY ITS TERMS AND CONDITIONS AND BY THE FOLLOWING:
CREDIT IS AVAILABLE BY NEGOTIATION OF YOUR DRAFT(S) IN DUPLICATE AT SIGHT FOR 100 PERCENT OF INVOICE VALUE DRAWN ON US ACCOMPANIED BY THE FOLLOWING DOCUMENTS:

1. SIGNED COMMERCIAL INVOICE IN 1 ORIGINAL AND 3 COPIES.

2. FULL SET 3/3 OCEAN BILLS OF LADING CONSIGNED TO THE ORDER OF THIRD HONG KONG BANK, HONG KONG NOTIFY APPLICANT AND

3. PACKING LIST IN 2 COPIES.

EVIDENCING SHIPMENT OF: 5000 PINE LOGS – WHOLE – 8 TO 12 FEET
FOB SAVANNAH, GEORGIA

SHIPMENT FROM: SAVANNAH, GEORGIA TO: HONG KONG
LATEST SHIPPING DATE: 02JUN97

PARTIAL SHIPMENTS NOT ALLOWED TRANSHIPMENT NOT ALLOWED

ALL BANKING CHARGES OUTSIDE HONG KONG ARE FOR BENEFICIARYS ACCOUNT.
DOCUMENTS MUST BE PRESENTED WITHIN 21 DAYS FROM B/L DATE.

AT THE REQUEST OF OUR CORRESPONDENT, WE CONFIRM THIS CREDIT AND ALSO ENGAGE WITH YOU THAT ALL DRAFTS DRAWN UNDER AND IN COMPLIANCE WITH THE TERMS OF THIS CREDIT WILL BE DULY HONORED BY US.

PLEASE EXAMINE THIS INSTRUMENT CAREFULLY. IF YOU ARE UNABLE TO COMPLY WITH THE TERMS OR CONDITIONS, PLEASE COMMUNICATE WITH YOUR BUYER TO ARRANGE FOR AN AMENDMENT.

信用狀範例 3（電報信用狀）：

Swift Field Descriptions

507 july 95 09:13　　page: 2355　　LP00
*** HARDCOPY msg id 0131-00010-00333 ***
RECEIVED FROM: IMPORTER'S COMMERCIAL BANK
TAIPEI, TAIWAN
sent to :
SELLER'S U.S. COMMERCIAL BANK
INTERNATIONAL DIVISION
SAN FRANCISCO, CA
date : 07 july 95 time : 09.13 issue of a documentary credit **urgent**
:27 /sequence of total :1/1

:40a/form of documentary credit	:IRREVOCABLE
:20 /documentary credit number	:DOC.500
:31C/date of issue	:950707 USA
:31D/date and place of expiry	:950921 USA
:50 /applicant	:IMPORTER'S COMPANY NAME IMPORTER'S COMPANY ADDRESS TAIWAN
:59 /beneficiary	:EXPORTER'S COMPANY NAME EXPORTER'S COMPANY ADDRESS USA
:32B/currency code amount	
currency code	: USD US DOLLAR
amount	: #100,000.00#
:39B/maximum credit amount	:NOT EXCEEDING
:41D/available with/by-name,address	:ANY BANK BY NEGOTIATION
:42C/drafts at	:SIGHT
:42D/drawee - name and address	:IMPORTER'S COMMERCIAL BANK
TAIWAN	
:43P/partial shipments	:PROHIBITED
:43T/transshipment	:PROHIBITED
:44A/on board/disp/taking charge	:USA PORT
:44B/for transportation to	:TAIWAN PORT
:44C/latest date of shipment	:950831
:45A/descr goods and/or services	:FUJI APPLES

CIF TAIWAN

:46B/documents required :+COMMERCIAL INVOICE AND THREE COPIES.

+FULL SET CLEAN ON BOARD BILLS OF LADING, MARKET FREIGHT

PREPAID CONSIGNED TO BUYER.

+INSURANCE CERTIFICATE.

+CERTIFICATE OF ORIGIN.

+USDA INSPECTION CERTIFICATE.

:47A/additional conditions :+ALL DRAFTS MUST INDICATE: DRAWN UNDER

IMPORTER'S

COMMERCIAL BANK TAIWAN LETTER OF CREDIT NUMBER DOC.500

:48 /period for presentation :DOCUMENTS ARE TO BE PRESENTED WITHIN 21 DAYS AFTER

SHIPMENT BUT WITHIN L/C VALIDITY.

:49 /confirmation instructions :WITH

:78 /instructions to pay/acc/neg bk:ALL REQUIRED DOCUMENTS ARE TO BE SENT TO IMPORTER'S

COMMERCIAL BANK, TAIPEI, TAIWAN IN ONE SET,

VIA COURIER

CONFIRMING THAT ALL TERMS AND CONDITIONS HAVE BEEN

COMPLIED WITH. DOCUMENTS ARE TO INCLUDE YOUR

SETTLEMENT INSTRUCTIONS.

:72 /sender to receiver information :THIS CREDIT IS SUBJECT TO THE UNIFORM CUSTOMS AND

PRACTICE FOR DOCUMENTARY CREDITS ICC PUBLICATION NO.

500, 1993 REVISION.

-AUT/**** Authentication Result

*END

Swift Field Descriptions

Most letters of credit are issued by electronic means. The following is a list of the fields in a SWIFT MT 700 message
(Issuance of Documentary Letter of Credit). Only a few fields are mandatory; most are optional and depend
on the nature of the transaction.
27 Sequence # (Page number within th4e total sequence)
40A Form of Documentary Credit (Irrevocable or Revocable)
20 Issuing bank’ s reference number
31C Date of issue
31D Date and place of expiry
51A/D Applicant bank/applicant reference number
50 Applicant
59 Beneficiary
32B Currency code and amount
39A Percentage credit amount tolerance
39B Maximum credit amount
39C Additional amounts covered
41A/B Available with (bank)...by (payment, negotiation, acceptance)
42C Drafts at (sight, time, etc.)
42A Drawn on (what party)
42M Mixed payment details (part sight, part time)
42P Deferred payment details
43P Partial shipments (allowed or prohibited)
43T Transshipments (allowed or prohibited)
44A Loading on board/dispatch/taking in charge from/at...
44B For transportation to...
44C Latest date of shipment
44D Shipment period
45A Description of goods and/or services
46A Documents required
47A Additional conditions
71B Charges (which party pays)
48 Period for presentation (within L/C validity)
49 Confirmation instructions (with/without)
53A Reimbursement bank
78 Instructions to paying/accepting/negotiating bank
57A “Advise Through” Bank
72 Sender to receiver information

範例 11-1　賣方：貨物正生產中，請快開信用狀

To jackturner@pioneer.com

From lindachen@tbi.com.tw

Cc

Bcc

Subject L/C Request

Attachment

Dear Jack,

We thank you very much for your order no. P12011 of March 3 together with your shipping instructions.

The goods are presently being manufactured. You informed us that you would arrange to issue an irrevocable L/C in our favor, valid until August 30. Please send the L/C promptly to avoid undue delay in shipment.

Upon arrival of the L/C, we will pack and ship the goods as soon as possible in accordance with your shipping instructions. We assure you that we will make complete shipment, and that you will be completely satisfied as to the quality.

Best regards,

Linda Chen
Sales Manager

翻譯

我們非常感謝您 3 月 3 日的訂單，號碼：P12011 以及出貨說明書。

我們目前正在生產您的商品。您有通知我們說您將開出以我方為受益人的不可撤銷信用狀，信用狀有效期至 8 月 30 日。請立即寄出信用證以避免出貨延遲。

我們一收到信用狀後，就會根據您的出貨說明書儘快包裝和運輸貨物。我們向您保證，我們會完成出貨事宜，並且確信您對我們的品質會完全滿意。

範例 11-2　賣方：貨物即將生產，請快開信用狀

To: jackturner@pioneer.com
From: lindachen@tbi.com.tw
Cc:
Bcc:
Subject: L/C for PO no. P12011
Attachment:

Dear Jack,

You confirmed your orders on April 8 and informed us that you had notified your bank to open an L/C. To date we have not yet received any information concerning your L/C from our bank.

Our production department has already prepared materials. Without your L/C, we cannot begin production and this places us under a lot of pressure since this order has been confirmed for a long time. So, please kindly check with your bank and advise us.

After receiving your L/C confirmation, we can arrange delivery of the first production samples within 15 days.

We look forward to hearing from you soon.

Best regards,

Linda Chen
Sales Manager

翻譯

感謝您 4 月 8 日的訂單，並告知我們您已通知您的銀行開立信用狀。但是，到目前為止我們還沒有收到從我方銀行通知貴公司有關信用狀的任何消息。

我們的生產部門已經準備好材料生產。如果沒有收到您的信用狀，我們就無法開始生產，這使我們面臨很大的壓力，因為這個訂單已經確認了一段時間。因此，煩請向您的銀行確認並通知我們。

在收到您的確認後，我們可以安排在 15 天內交付第一批生產的樣品。

我們期待您的佳音。

範例 11-3　賣方：出貨前跟催 T/T 付款

To	jackturner@pioneer.com
From	lindachen@tbi.com.tw
Cc	
Bcc	
Subject	T/T for PO no. P12011
Attachment	

Dear Jack,

Regarding your order no. P12011, we are pleased to inform you that it will be ready for shipment around Nov. 21. We have tried to complete it on schedule. Please remit the payment (totally US$20,000) by T/T at your earliest convenience and on receipt of your remittance, we will effect the shipment without any delay.

We look forward to the confirmation of your remittance in the near future.

Best regards,

Linda Chen
Sales Manager

翻譯

關於你的訂單號碼 P12011，我們很高興地通知您，它將在 11 月 21 日左右出貨。我們將依訂單要求完成並出貨。請以電匯方式付款，總金額 20,000 美元，我們將在收到款項後立即出貨。

希望能立刻收到您匯款的確認。

常用句

1. We wait for an early arrival of the L/C.
2. Please open an irrevocable letter of credit in our favor.
3. The time of shipment is fast approaching and we must ask you to send the L/C and ship instructions immediately.
4. We will ship your goods by the first available ship upon receipt of an L/C.
5. We will ship your samples after we receive your T/T payment.
6. We shipped your goods last Friday, as agreed payment terms, please make T/T payment in 30 days.

練習題

1. 將下列的句子翻成英文

A. 請以本公司為受益人開出不可撤銷的信用狀。

B. 若貴公司儘速開出信用狀，本公司將極為感謝，以便照合約規定出貨。

C. 茲收到貴公司開出的金額 USD50,000 的信用狀，謝謝。

D. 請您修改信用狀。

E. 請延長信用狀的最後期限與交貨日期至 8 月 30 日。

F. 請延長出貨期限至 7 月底，或是修改信用狀，以便轉運。

2. 將下列中文書信翻譯成英文

有關貴公司訂單號碼 PO AA254，我們很高興通知您貨物已準備好可以立即出貨。但是我們尚未收到貴公司的信用狀。

請告知是否已開出信用狀，如已開出請將信用狀的明細告知或是傳真信用狀影本給我們，一收傳真，我們將立刻安排出貨。

3. 請寫一封跟催信用狀的信函。

4. 請寫一封出貨前跟催電匯的信函。

5. 請寫一封信函通知客戶修改信用狀裝船期自 8/10 延至 8/30，有效期自 9/10 延至 9/30，以符合貨品出口日期，請客戶儘速修改並告知。

6. 請寫一封信感謝客戶我們已收到信用狀，號碼：880128，但是我們發現信用狀上有 3 點錯誤，(1) 價格條件應該是 FOB Twiwan(2) 應該要允許轉運 (3) 最後出貨日應為 10 月 30 日。以上請儘快修改並確認。

CHAPTER

12 出貨

Shipment

出口商在商品生產完工後，即根據訂單上的要求安排出貨事宜。一般說來如無特殊要求通常都是以海運為主，買賣雙方在出貨前也會有一些協調事項，例如出貨包裝有無特殊規定、空運要雇用哪一家航空公司或是海運要用哪一家的航運公司的船等。

一般而言，在 FOB 的條件下，海運或是空運事項均由買方處理。因此賣方在確定船期後，可先通知客戶，讓客戶了解預定出貨的情形。客戶在收到賣方的出貨通知時即可根據此預先的出貨通知辦理保險、安排倉儲等事宜。

出貨後，賣方依據出貨內容備齊相關文件（如 Invoice，Packing List，B/L…等），這些文件可先傳真或是 e-mail 給客戶，以便先行核對以利客戶儘速取得商品。客戶能否順利提取商品與出貨文件有很大關係，故賣方應儘早將正本文件寄給客戶。

賣方如果於訂單規定的時間內無法如期出貨，應儘快告知客戶延遲出貨的日期，取得客戶同意後，應檢視信用狀上條款有無違規，如果有違規事項，應請求客戶修改信用狀條款。

出貨通知內容上應包含下列幾項：

- 訂單號碼
- 船名 (name of vessel) 或是班機號碼 (flight no.)
- 預計離港日 (ETD)
- 預計到港日 (ETA)
- 提單號碼（B/L No. 海運）或是（AWB No. 空運）

範例 12-1　賣方：通知出貨事宜

To ▷ jackturner@pioneer.com

From ▷ lindachen@tbi.com.tw

Cc ▷

Bcc ▷

Subject ▷ PO #AA1245 Shipment

Attachment ▷

Dear Jack,

We are pleased to confirm that your PO No. AA1245 is now ready for delivery.

When placing your order you stressed the importance of prompt delivery by air shipment, and I am glad to say that by making a special effort we have been able to improve by a few days on the delivery date agreed.

Please let us have your shipping instructions, and as soon as we hear from you we will send you our advice of dispatch.

Best regards,

Linda Chen
Sales Manager

翻譯

我們很高興通知您，您的訂單編號 AA1245 現已準備好出貨。

您在下單時特別強調要儘早用空運出貨，我很高興地說，經過特別的努力，我們可以提前幾日交貨。

請儘快告知出貨有無任何要注意的地方，一旦確認後，我們出完貨會立即告知出貨的明細。

範例 12-2　賣方：準備出貨通知

To	jackturner@pioneer.com
From	lindachen@tbi.com.tw
Cc	
Bcc	
Subject	PO #AA1245 Shipment
Attachment	

Dear Jack,

We are pleased to confirm that the goods you ordered in your order AA1245 dated on April 3rd are packed and ready for despatch. The consignment consists of two cases, each weighing about 100 kg.

We have made arrangements for shipment, CIF L.A., with Watsons, your forwarding agent. We will arrange for shipping documents to be sent to you through Bank of America against our draft for acceptance, as agreed.

We look forward to further business with you.

Best regards,

Linda Chen
Sales Manager

翻譯

我們很高興地通知您，您在 4 月 3 日訂購的訂單 AA1245 中所訂購的商品已包裝好並可隨時出貨。這張訂單包含兩個箱子，每個箱子重約 100 公斤。

我們已經聯絡貴公司在洛杉磯的攬貨公司 (Watsons) 做了安排。我們將依雙方約定的付款方式，透過美國銀行提示出口文件，以承兌交單的方式請求付款。

我們期待未來與您有進一步的業務往來。

範例 12-3　賣方：海運出貨通知

To jackturner@pioneer.com

From lindachen@tbi.com.tw

Cc

Bcc

Subject Shipping details

Attachment invoice.doc; packing.doc; B/L.doc

Dear Jack,

We are glad to inform you that your order #AA1245 has been finished and we have booked a space for Oct. 8 through your appointed forwarder-GEODIS. The shipping details are listed as below:

1. Closing Date: Oct. 5, 2021
2. ETD: Oct. 8, 2021
3. ETA: Nov. 4, 2021
4. Vessel name: Santa Maria V.0032

Attached please find the copies of B/L, Invoice and Packing list. The original documents will be sent today by FedEx.

Please let me know if you have any questions.

Best regards,

Linda Chen
Sales Manager

1. forwarder：攬貨公司
2. ETD：Estimated Time of Departure 預計開航日
3. ETA：Estimated Time of Arrival 預計到達日
4. B/L：Bill of Lading 海運提單
5. FedEx：Federal Express 美國快遞公司

翻譯

我們很高興地通知您，貴公司訂單 # AA1245 已完成，我們已經透過貴公司指定的攬貨公司 GEODIS 在 10 月 8 日訂了船班出貨。出貨的詳細資料如下：

1. 結關日：2021 年 10 月 5 日
2. 預計開船日：2021 年 10 月 8 日
3. 預計到達日：2021 年 11 月 4 日
4. 船名：Santa Maria V.0032

附件是提單，發票和裝箱明細表。我們將原始出貨文件於今日由聯邦快遞寄送給您。

如果您有任何問題，請來函告知。

範例 12-4　賣方：空運出貨通知

To	jackturner@pioneer.com
From	lindachen@tbi.com.tw
Cc	
Bcc	
Subject	Shipping advice (or shipping notice)
Attachment	

Dear Jack,

We have received your instruction to ship PO #AA1245 by air. As requested, we have delivered the goods to your air freight forwarding agent. The flight details are as follows:

MAWB: 160HKG7684 4854
HAWB: HKG0173 1693
Air Line: Cathay Pacific Airways CX-082 (May 20)

We sent you the original documents by UPS to your mailing address. Please let me know if you have any questions.

Best regards,

Linda Chen
Sales Manager

1. MAWB：主空運提單 (Master Air Waybill)
2. HAWB：空運大提單 (House Air Waybill)

翻譯

我們已收到您的出貨指示，指示我們以空運運送訂單 AA1245。我們已依您的指示，透過貴公司的空運攬貨公司安排出貨。出貨航班詳情如下：

主空運提單：160HKG7684 4854
空運大提單：HKG0173 1693
航空公司：國泰航空 CX-082（5 月 20 日）

我們原始出貨文件以 UPS 寄到您的郵寄地址。如有任何問題，請告知。

範例 12-5　賣方：出貨延期，請客戶修改信用狀

To	jackturner@pioneer.com
From	lindachen@tbi.com.tw
Cc	
Bcc	
Subject	Delay shipment for PO #AA1245
Attachment	

Dear Jack,

We are sorry to inform you that the delivery for PO #AA1245 will be delayed until end of July, due to a typhoon that has recently hit Taiwan. We have advised you the whole situation suffered from our factory two days ago.

We are sorry for this unfortunate delay. However, it is indeed a case of force majeure, which is beyond our control. Because of this delay, we are unable to ship the goods before the deadline shown on the L/C. Therefore, we have to ask a favor of you to amend your L/C at once. Please amend the shipping date and expiry date to July 30 and August 15 respectively.

With our best efforts, we guarantee to deliver the goods by the first direct vessel and confirm the shipping details further. We hope it won't cause you too much inconvenience.

Your kind understanding to the above is highly appreciated. We look forward to your prompt cooperation in amending your L/C soon.

Best regards,

Linda Chen
Sales Manager

註
1. force majeure：不可抗力因素
2. amend L/C：修改信用狀

翻譯

我們很抱歉地通知您，由於颱風最近襲擊台灣，您的訂單 AA1245 的出貨日期將延期至 7 月底。兩天前我們已告知您我們工廠這一次受颱風影響的狀況。

我們對這次的延誤感到抱歉。但是，這確實是不可抗拒的因素而造成，非我們所能控制的。因為這次的出貨延遲使我們無法在信用狀的截止日期前裝運出貨。因此，我們要請您立即修改信用狀。請將發貨日期和有效日期分別修改為 7 月 30 日和 8 月 15 日。

我們將盡最大的努力並保證我們會用第一艘直達船來交付貨物並通知您出貨明細。希望這一次不會給您造成太多不便。

感謝您對上述的理解。期待您儘快修改信用狀。

常用句

1. We are pleased to say that these goods have been dispatched today.
2. We hope the goods reach you safely and that you will be pleased with them.
3. We are pleased to inform you that your order was shipped on May 10 by M/S Tao Yuan leaving Taichung for Hong Kong.
4. We have forwarded your order by the first available express cargo airplane.
5. The Chung Hsin sailed on July 10 from Kaohsiung bound for Los Angeles.
6. We will send a set of the shipping documents covering your order no. xxx separately.
7. As requested, the triplicate commercial invoice and duplicate inspection are enclosed.
8. Your order was airfreighted today with shipping documents airmailed.
9. There is no ship to your port during March.
10. We see so possibility of shipping the goods on or before April 10.
11. We have to advise you that we are unable to dispatch in full your order owing to a great shortage of shipping space.

練習題

1. 將下列的句子翻成英文

A. 本公司已訂好在 11 月 3 日左右從基隆出航的「中國第一」的船位。

B. 已向漢陽輪預定好船位了，大約 11 月 3 日離開高雄。

C. 您的貨物已經準備好可以出貨了，很樂意收到您立即的指示。

D. 我們向您確認 100 台相機將由 CAL520 班機送抵紐約。

E. 收到您的信用狀後就可以立即出貨。

F. 很高興通知您貨物今早已轉由陸運送到芝加哥了。

G. 相信商品會平安抵達您的手中，而您也會很快再下單。

H. 根據您的要求，寄上 3 份商業發票和 2 份檢驗證明。

I. 附上包裝明細、提單、發票、產地證明及檢驗報告。

J. 您的貨物已空運，裝貨文件亦已用航空信件寄出。

K. 我們的承運公司會將裝船文件寄給您。

L. 下列文件已經由台灣銀行轉送紐約的美國銀行。

M. 我方無法在您要求的日期內出貨。

N. 由於年關已近，港口擁塞，本批出貨無法將您所訂的商品出完。

O. 目前我們敢承諾的最早交期是明年 1 月。目前船員罷工，恐怕無法如期交貨。

P. 8 月中均沒有船前往您的港口。

2. **將下列中文書信翻譯成英文**

有關訂單號碼 PO123，我們很高興通知貴公司我們已於今日出貨 100 箱的女性運動鞋，經長榮海運公司，船名 "EVA V-456"，預計離開基隆港日期為 8/12，預計抵達舊金山為 8/20。

我們已於今日用快遞方式寄出一套副本出貨文件提供您參考。

我們相信貨物會完好抵達，謝謝惠顧。

3. **請寫一封信告知客戶，貴公司訂單 PO123 已於今日裝船。船名為總統號，預定 8/8 由台中港駛向紐約港，到期日預計為 8/25。**

4. **請寫一封信告知客戶因地震的關係，我們無法以訂單原訂出貨日出貨，會延期出貨日至 10 月底，並請客戶修改信用狀之出貨日與到期日，一但確認出貨船期將立即告知。**

5. **請寫一封空運出貨的通知函，告知客戶已遵其指示，將貨物一半空運，一半海運。空運主要是把航班，主提單 (MAWB) 及空運公司發的提單 (HAWB) 一併給客戶。海運會在確認船班後告知詳細資料。原始出貨文件將以 UPS 寄送。**

CHAPTER

13 催款

Collection

一筆交易進行至賣方將貨物出貨後，文件寄給買方，買方即應於雙方同意的付款條件下將貨款交付買方。但是不同的付款方式將有不同的處理方式，以下就不同的付款方式賣方可能碰到的情況加以說明。一般說來，如果客戶於出貨前即償付貨款，賣方就無需催款。但是如果雙方的條件是出貨後買方才償付貨款，這時賣方就須適時跟催款項（出貨前的 T/T 與 L/C 跟催請參照 Chapter 11）。

很不幸地，有時候就是會出現客戶拖欠款項的狀況，也許是客戶的問題，也許是有其他因素而造成客戶無法及時償付或款。如果出現這種狀況，賣方就需要立即通知客戶，書信中須維持一貫的禮貌，但是態度要堅定。

催款信的語氣應該視債務的積欠時間、是否習慣性拖欠、以及客戶的重要程度等因素而定，必須個別處理每個個案。碰上棘手的狀況時，接受對方支付的部分付款會比耗費金錢與時間訴諸法律行動來的好，法律行動是最後的手段。但無論如何，每一封信都應保持一貫的禮貌與堅定的態度。

催款信的撰寫應該注意下列事項：

1. **寫給對應窗口 (contact window)**：商品出貨後，款項的催收是銷售人員 (sales) 的責任，因此銷售人員要聯繫客戶端的往來窗口（採購人員／ buyer）進行貨款的催收。

2. **帳務要明確**：銷售人員應將客戶未付款項目提出訂單號碼、出貨明細、發票號碼等資料列出，以方便客戶找到相應的紀錄。提醒客戶雙方訂定的付款條件，對於延遲的帳目，你希望客戶如何付款。

3. **態度中立**：催款信中口氣應該持平，僅敘述事實的真相，切不可有責怪客戶的口氣。客戶可能出於各種原因而拖延付款，我們只要保持一貫的服務態度即可。

4. **請對方付款**：寄催款信給客戶時，客戶如果來函說明拖欠款項的原因，或是對你的催款信有所回應，應該讓客戶有機會進行解釋。這樣的舉動顯示你關心客戶，比較容易在催款過程中達到雙方可接受的付款方式。

5. **謝謝客戶**：即使客戶支付款項的時間晚了，仍舊要謝謝客戶。

13.1 出貨後的貨款跟催

買賣雙方的付款約定條件如果是承兌交單 (D/A)，或是付款交單 (D/P)，則賣方於出貨後就應立刻發函給買方跟催貨款。

範例 13-1　賣方：通知客戶我方已出貨，已開出匯票請其承兌 (D/A)

To: jackturner@pioneer.com
From: lindachen@tbi.com.tw
Cc: judywong@fcb.com.tw
Bcc:
Subject: Request payment
Attachment: 4 files

Dear Jack,

Please be informed that we have shipped your order #9923 today by the vessel: "May Flower V-237" from Keelung to New York, ETD: Sep. 20, 2021, ETA: Oct. 25, 2021. We believe that it will reach you in good condition.

Attached please find shipping documents, we also sent following documents to you by separate mail today.

Commercial Invoice: 2 copies
Packing List: 2 copies
Bill of Lading: 1 copy
Certificate of Origin: 1 copy

We have drawn on you a 60-day draft for US$15,000 through our banker, the First Commercial Bank. The original shipping documents, along with the draft, will be handed to you by our banker through the Bank of America on your acceptance of the draft.

We shall be glad if you will duly protect the draft on presentation, and we look forward to your repeat order confirmation in the near future.

Best regards,

Linda Chen
Sales Manager

翻譯

請注意，您的訂單 9923 已於今日經由船班 :May Flower V-237 從基隆港出貨到紐約，預計開船日：2021 年 9 月 20 日，預計到達日：2021 年 10 月 25 日。相信貨到時的狀況會非常良好。

附件是出貨單據，另外我們還於今天以郵寄方式將下列文件寄給您：

商業發票：2 份
包裝明細表：2 份
提單：1 份
產地證明書：1 份

我們已透過我們的銀行－第一商業銀行－簽發了一張 15,000 美元的 60 天匯票。我們的銀行原始出貨文件連同這張匯票經由美國銀行提示給您。

感謝您對匯票的承兌，期待在不久的將來再次收到您的訂單。

範例 13-2　賣方：通知客戶我方已出貨，已開出匯票請其付款 (D/P)

To	jackturner@pioneer.com
From	lindachen@tbi.com.tw
Cc	judywong@fcb.com.tw
Bcc	
Subject	Request payment
Attachment	3 files

Dear Jack,

Please note that your P/O #9956 has been shipped per "Maersk steamer "Sally Maersk V.900W", sailing on Mar. 5, 2021 ETA London Apr. 9, 2021. Enclosed please find the invoice, packing list and B/L for your reference.

To cover this shipment, we are drawing a draft on you at sight, D/P, for US$7,800 and we are asking you to protect it. The shipping documents will be sent through the Bank of New York against your payment of our draft. Please protect the bill without delay.

Thank you for your attention to the above.

Best regards,

Linda Chen
Sales Manager

請注意，您的訂單 9956 已經由船班 "Sally Maersk V.900W" 於 2021 年 3 月 5 日起航，預定到達倫敦的日期是 2021 年 4 月 9 日。隨函附上發票，裝箱明細和提單供您參考。

本公司已開立了一張美元 7,800 的匯票，請您向指定的銀行付款。出貨文件將經由紐約銀行給您並請求付款。請勿延遲付款。

感謝您對上述的關注。

13.2 逾期未償付貨款的跟催

撰寫催款信的困難就在於在催款的同時，依舊要保持客戶至上的服務態度。信中的態度讓客戶覺得禮貌得體非常重要，畢竟你並不知道為什麼客戶會沒付款，也不知道到底是不是客戶本身的錯誤，一切尚待詢問清楚。因此第一封的催款信，語氣依舊要有禮貌，不可責怪客戶。先將這個狀況視為某種出錯或是疏忽所造成，請求客戶立即支付逾期貨款。

如果客戶對第一封的催款信沒有任何回應，我們就要假設客戶可能面臨財務問題，有可能拖欠貨款，或是不付款的現象。此時，我們要持續追蹤並且態度也應該要越來越強硬，客氣的成分越來越少，並且將信件的附件要呈報上級與相關單位，例如財務部、法務部等。對於未付款項，公司可能已有固定的政策。有時候，恰當的做法是提供客戶新的付款條件，比如說在某一個期間內付款可享現金折扣、分期付款或是延後付款期限等方法。至於該採取何種做法，取決於公司的政策，以及公司與客戶之間的關係。對於財務有風險的客戶，銷售人員應主動告知上級主管，由公司決定於貨款未收到前，是否對其後續的訂單繼續出貨。

範例 13-3　賣方：匯票到期未兌付

To jackturner@pioneer.com
From lindachen@tbi.com.tw
Cc judywong@fcb.com.tw
Bcc
Subject Request payment
Attachment

Dear Jack,

We got a letter of advice from our bank that our draft No. 1123 for US$9,000, due on July 25, against Invoice #KK-856, has not been paid yet. We are writing you immediately and asking you to take it as an urgent matter.

Please kindly honor our draft within one week after receiving this e-mail. Otherwise, we will be forced to take some legal actions. Your urgent cooperation in this respect will be highly appreciated.

We look forward to your confirmation soon.

Best regards,

Linda Chen
Sales Manager

翻譯

我們的銀行通知我們，我們匯票號碼 1123，金額美元 9,000，到期日是 7 月 25 日（對應發票號碼 :KK-856）還沒有支付。請立刻處理此一緊急事項。

請在收到此電子郵件後的一周內儘快付款。否則，我們將被迫採取一些法律行動。感謝您立即的協助。

期待能儘快收到您的確認。

範例 13-4　賣方：帳款到期未兌付

To	jackturner@pioneer.com
From	lindachen@tbi.com.tw
Cc	judywong@fcb.com.tw
Bcc	
Subject	Request payment
Attachment	collection statement

Dear Jack,

As we always receive your payment punctually, we are wondering to have had neither remittance nor response in connection with our invoice #9928.

As the settlement for this invoice is 5 weeks overdue, we think you may not have received our statement. We have resent a copy as per attached and we hope it will receive your fullest attention and make prompt payment.

Best regards,

Linda Chen
Sales Manager

翻譯

一直已來您的付款從來沒有延遲過，所以我們對於您尚未支付發票 9928 的款項感到奇怪。

這一張發票的貨款已逾期 5 週，我們猜想您可能未收到我們的對帳單。所以我們重新再寄一次副本發票給您，希望您收到後迅速付款。

範例 13-5　賣方：帳款到期未兌付

To jackturner@pioneer.com
From lindachen@tbi.com.tw
Cc
Bcc
Subject Settlement of account
Attachment 1 file

Dear Jack,

According to our records, I want to remind you that your order #5566 shipped on June 20 from Taiwan has not been paid. I have enclosed a detailed statement for this amount of US$11,000 and hope you can make a prompt settlement.

Please kindly arrange it by T/T without delay and advise. We thank you in advance.

Best regards,

Linda Chen
Sales Manager

翻譯

根據我們的記錄，我想提醒您，您的訂單 5566 已於 6 月 20 日從台灣出貨，但是到目前為止訂單貨款尚未支付。附件為對帳明細，總額為 11,000 美元，希望您能及時付款。

請立即以電匯償付貨款並告知。在此先感謝你。

範例 13-6 賣方：帳款到期未兌付經催繳仍無回應 I

To jackturner@pioneer.com

From lindachen@tbi.com.tw

Cc

Bcc

Subject Payment overdue

Attachment

Dear Jack,

According to our records, your account has US$45,000 overdue. Following are the details.

P/O No.	Shipping Date	Overdue Balance
PO2564	May 15	US$15,000
PO8789	May 25	US$ 9,000
PO9921	June 11	US$21,000
Total:		US$45,000

If you have settled the account before this letter reaches you, please ignore this reminder. Thank you for your payment and your business. As always, please contact me if you have any questions or concerns.

Best regards,

Linda Chen
Sales Manager

翻譯

根據我們的帳上紀錄，您的帳戶餘額以有 45,000 美元到期未支付。以下是帳目明細：

訂單號碼	出貨日	未償付貨款餘額
PO2564	5 月 15 日	US$15,000
PO8789	5 月 25 日	US$ 9,000
PO9921	6 月 11 日	US$21,000
總額：		US$45,000

假如您於收到本信函之前已支付貨款，請不用理會本函通知。感謝您的付款與訂單。如有任何問題，歡迎來函詢問。

範例 13-7　賣方：帳款到期未兌付經催繳仍無回應 II

To jackturner@pioneer.com
From lindachen@tbi.com.tw
Cc
Bcc
Subject Settlement of account
Attachment

Dear Jack,

Please kindly note that our statement of account dated April 6 is still waiting for settlement. As your payment is now overdue over 2 months, we must ask you to remit the outstanding balance immediately. Please make payment without any delay.

Thank you for your payment and your business.

Best regards,

Linda Chen
Sales Manager

翻譯

請注意，我們 4 月 6 日的帳戶明細記錄您尚未支付貨款。由於您的貨款已超過 2 個月未支付，因此我們必須要求您立即償付貨款。請立即付款。

感謝您的付款和訂單。

範例 13-8　賣方：帳款到期未兌付經催繳仍無回應 III

To: jackturner@pioneer.com
From: lindachen@tbi.com.tw
Cc:
Bcc:
Subject: Settlement of account
Attachment:

Dear Jack,

Your balance on account no. 7721 is overdue for 45 days and stands at US$6,000. This is a second reminder, please send payment on receipt of this notice.

We would like to continue to serve you on open account terms, but to do this, your bills must be paid within 30 days as agreed. As soon as the overdue sum is cleared, we will fulfill your current order.

Thanks you for your cooperation.

Best regards,

Linda Chen
Sales Manager

翻譯

您的帳戶號碼 7721 已有未支付貨款餘額 6,000 美元，已逾期 45 日。本信函是第二次未付款通知，請收到本通知後立即付款。

我們希望繼續提供您記帳的交易模式，但是為維持此一條件您必須如雙方同意的條款於 30 日內償付貨款。逾期款項一經支付，我們就會立刻進行您目前的訂單。

感謝您的合作。

範例 13-9　賣方：多次跟催無回應 I

To	jackturner@pioneer.com
From	lindachen@tbi.com.tw
Cc	
Bcc	
Subject	Final notice-payment overdue
Attachment	

Dear Jack,

Your balance of $8,835.23 is now more than 60 days overdue. We have written to request payment over three times but I am afraid we still have not heard from you in reply.

Please kindly arrange the payment and let us receive it by the end of September. Otherwise we will have to take legal action.

Thank you for your prompt attention to this matter.

Best regards,

Linda Chen
Sales Manager

翻譯

您的帳上餘額 8,835.23 美元已超過 60 天尚未支付。我們已經寫過三次的信件去通知您付款，但是我們一直沒收到您的回覆。

請立刻安排付款，讓我們在 9 月底前能收到。否則，我們將不得不採取法律行動。

感謝您及時關注此事。

範例 13-10　賣方：多次跟催無回應 II

To: jackturner@pioneer.com
From: lindachen@tbi.com.tw
Cc:
Bcc:
Subject: Final notice-payment overdue
Attachment:

Dear Jack,

Attached is the account statement of July. You will see that your account has been outstanding for a long time and has now become delinquent.

We have written to you on three occasions and have not received a reply. We cannot understand why you have no action to settle your account. We regret to inform you that unless we receive full payment within ten days, your account will be cancelled and your credit will be terminated.

We look forward to an early reply.

Best regards,

Linda Chen
Sales Manager

翻譯

本信函附上 7 月份的帳戶明細表。你會發現你的賬戶已經很長時間沒支付貨款了，現在已經變成了積欠款。

我們已經三次給你寫信，但是一直沒有收到您的答覆。我們無法理解您為何沒有償付貨款。在此我們很遺憾地通知您，除非我們在 10 天內能收到全額付款，否則您的賬戶將被取消，您的信用也將被終止。

我們期待您早日的回覆。

範例 13-11　賣方：多次跟催無回應 III

To: jackturner@pioneer.com
From: lindachen@tbi.com.tw
Cc:
Bcc:
Subject: Final notice-payment overdue
Attachment:

Dear Jack,

We regret not to have response from you since we have sent at least 3 mails to remind you of the balance $8,825.35 still unpaid.

This is all the more disappointing because of our past good relationship over many years. Unless the amount due is paid or a satisfactory explanation received by the end of this week, we shall be reluctantly compelled to put this matter to our lawyers.

We look forward to hearing from you soon.

Best regards,

Linda Chen
Sales Manager

翻譯

我們遺憾在我們寄了至少 3 封的郵件通知您償付未支付貨款，金額 8,825.35 美元之後，一直都沒收到您的回覆。

由於我們過去多年的良好關係，這樣的情形更令人失望。除非我們於本週末能收到您支付的貨款或收到令人滿意的解釋，否則我們將不得不將這件事情交給我們的律師處理。

期待您的回音。

範例 13-12　賣方：多次跟催無回應 IV

To jackturner@pioneer.com
From lindachen@tbi.com.tw
Cc
Bcc
Subject Final notice-payment overdue
Attachment

Dear Jack,

We have written to request payment for following overdue balance many times. However, we regret that we have not received your full payment yet.

P/O No.	Shipping Date	Overdue Balance
PO2564	May 15	US$15,000
PO8789	May 25	US$ 9,000
PO9921	June 11	US$21,000
Total:		US$45,000

Please understand it's a serious situation for our financial department. As a salesperson, I am experiencing much pressure about this matter. I do believe your credit and reputation are still good. Therefore, please kindly try your best to settle this matter urgently. If we do not receive your payment by end of this month, we will have to turn the account over to our lawyers.

Your prompt reply and early settlement will be needed and appreciated.

Best regards,

Linda Chen
Sales Manager

若客戶對第一次發出的催款信都沒有回應，賣方須持續催，直到收到貨款。在跟催信函中客戶若是提出不同的理由拖延付款，此時銷售人員須提高警覺是否客戶端已有財務危機，須及時呈報上級作進一步處理。

翻譯

我們已寫過多次信函要求您償付以下未支付的貨款。但是，很遺憾的是我們一直沒收到您的付款。

訂單號碼	出貨日	未償付貨款餘額
PO2564	5 月 15 日	US$15,000
PO8789	5 月 25 日	US$ 9,000
PO9921	6 月 11 日	US$21,000
總額：		US$45,000

請理解這對我們財務部門來說是非常嚴重的情況。作為一名銷售人員，這件事情讓我感到很大的壓力。我相信您的信用和聲譽仍然很好。因此，請儘快處理此事。如果我們在本月底沒有收到您的付款，我們將不得不將帳戶轉交給我們的律師來處理。

感謝您立刻的回答和付款。

常用句

1. Please pay the balance to our bank, the First Bank, on the 15th of each month.
2. In order to cover the shipment, we have drawn today on you at sight and ask you to pay our bill on presentation.
3. We shall be much obliged if you will give a little more time to settle the account due.
4. We have not received your L/C to cover your order No. XXX.
5. According to our conditions of sales, your remittance was due on July 31.
6. According to our record, your account balance USD10,000 was overdue for 2 weeks.
7. We have to insist upon full payment of your account.
8. Your failure to pay on time is in turn causing us financial embarrassment.
9. We insist upon full payment of your account, otherwise, we shall be forced to take legal action.
10. We must insist on receiving payment by March 3rd, if not, we shall be compelled to take legal action.
11. An early remittance will be appreciated.
12. Please confirm when we can expect the payment.
13. We urge that you make this settlement immediately. Your cooperation is appreciated.

練習題

1. 將下列的句子翻成英文

A. 請於每月 15 日將貨款匯入我們在第一銀行的戶頭。

B. 請於本月底前寄貨款給我們。

C. 提醒您的貨款於 5 月 10 起已經過期了。

D. 我們堅持要全額付款，否則將採取法律行動。

E. 我們必須堅持於 3 月底前收到貨款，否則我們將訴諸法律行動。

F. 請指示貴方銀行於一星期內匯款給我們。

G. 催促您不要拖延，趕緊匯款。

H. 我們了解您的立場，但是我們幾乎已經無法承受您匯款的再三拖延。

I. 我們一直提醒您尾款未付，卻一直未收到回音或匯款。

J. 您無法準時付款，導致我方財務困難。

K. 根據本公司的紀錄，貴公司還未償付訂單 AA-123 的款項。

L. 如果沒有立即收到款項，我們就必須通知我們的律師了。

M. 如果付款事宜已在進行中，在此謝謝您。

2. 將下列中文書信翻譯成英文

我們在此附上對帳單號碼 56412，顯示本月份到期的貨款總額為 US$13,365.78。

請儘快用電匯付清欠款。謝謝惠顧並等候您的通知付款狀態。

3. 根據下列中文情境，寫一封催款信

有關貴公司的本月應償付款項，我們在此附上應收帳單 88-123，金額總共為 US$8,542。由於貨物已於上個月初裝船出貨，我們希望能在這個月底前收到貨款。請於匯款同時發 e-mail 通知我們。

4. 根據第 3 題，寫一封第二次的催款信。

5. 請寫一封催款信告知客戶，我方已對客戶積欠 2 個月的貨款 US$17,500 提醒過至少 2 次，但是都未收到客戶的回應。請客戶須立即清償帳款，否則本公司將交由公司的律師處理此事。

6. 請寫一封催款信表示我們已於 3 月 10 日即 3 月 23 日去函提醒客戶 2 月份的對帳單欠款餘額 US$6,800.23 的事情，至今尚未收到任何回覆，令人感到失望。因為彼此過去幾年的合作關係良好，所以我們更感失望。在此情況下，除非客戶於 10 天內回覆，否則我們必須認真考慮進一步的行動，以取回欠款。

7. 請寫一封最後通牒的催款信，表示我方已於 7 月 15 日再次去函通知客戶逾期已久的積欠金額 US$1,534.03 的事情，很驚訝至今仍未收到回覆。過去雙方合作關係一直很好，即使如此，我們仍不允許帳款無限期不繳付。除非客戶在本月底前結清，或是提出令人滿意的解釋，否則我們將被迫交由我們的律師處理此事件。

CHAPTER

14 抱怨信

Complaint

在國際商業交易的活動中，若商品平安抵達買方手中，並且買方滿意賣方的商品與服務，則此交易於買方順利收取貨款後，即視為交易完成並告一段落。但是有時候買方在剛收到商品時或是收到商品後，對賣方所提供的商品、服務或其他的商務運作方式有不滿意之處，此時就會有抱怨 (complaint) 的產生。嚴重時，則買方會根據雙方買賣合約的條款提出索賠 (claim)。

客戶常抱怨或是提出索賠的原因，大概可以列為下列幾項：

1. 貨品損壞 (damaged goods)
2. 出錯貨品 (wrong goods)
3. 品質不良／瑕疵 (quality issue/defective goods)
4. 數量短缺 (shortage)
5. 尺寸或是規格不符 (specification issue)
6. 包裝不良 (poor package)
7. 出貨延遲 (delayed delivery)
8. 裝船延誤 (delayed shipment)
9. 文件錯誤 (wrong documents)
10. 違反合約 (against contract)
11. 取消合約 (cancel contract)

14.1 如何撰寫抱怨信

買方在寫抱怨信卜時必須在「態度堅決」與「維持理性」之間找到平衡。你要清楚表達不滿之處，但是不需要把供應商當成敵人看待，供應商與你應該是合作的關係。

如果不滿意所得到的商品或是服務，應該給供應商解決問題的機會。如果有必要寫抱怨信，請告訴對方你希望的解決方式，並給對方一個期限。在撰寫抱怨信的時候，應該注意幾個重點：

1. 精確具體：確切說明問題，並提供相關文件讓對方可以追查。
2. 提出明確要求：清楚描述你的要求，並說明本身遭受的損失與不便，要求對方在一定的合理時間內回覆。

3. 公正合理：所提的要求應符合公正合理，如此供應商才會有解決的意願。

一般而言，很多公司都設有一定的處理客戶抱怨的程序，依照對方的程序去進行抱怨投訴，可以節省不必要的時間浪費。當然，撰寫抱怨信應該條理分明，維持禮貌。威脅或是譏諷都不會帶來任何好處，要保持專業與禮貌的語氣。因為寫信的目的在謀求問題的解決，而不是造成雙方關係的緊張。同時，對不良情況的描述，不可無中生有或是誇大其詞。如果寫完抱怨信後，對方都沒有回應，就需再寄一封語氣更強烈的抱怨信去跟催，或是把信寄給更高層的主管。

範例 14-1　買方：抱怨數量短缺

To	lindachen@tbi.com.tw
From	jackturner@pioneer.com
Cc	
Bcc	
Subject	Quantity shortage for PO#1332
Attachment	

Dear Linda,

We received PO#1332 yesterday. However, on opening the container we found that the goods were short by 20 units of electric drills. We need this goods very urgently as we have very little stock. We ask you ship the 20 drills by air parcel immediately.

Let me know the air shipment details when available.

Regards,

Jack Turner
Purchasing Manager

翻譯

我們昨天收到了訂單 1332 的商品。但是，在打開貨櫃時，我們發現貨物短缺了 20 台電鑽。我們急需這些商品，因為我們目前的庫存很少。要請您立即通過航空包裹運送這短少的 20 台。

請告知空運詳情。

範例 14-2　買方：抱怨品質不佳 I

To: lindachen@tbi.com.tw
From: jackturner@pioneer.com
Cc:
Bcc:
Subject: Quality issue for pens
Attachment:

Dear Linda,

We have recently received several complaints from customers about your pens. The pens are clearly not satisfactory, and in some cases we have had to refund the purchase price.

The pens are part of the batch of 500 supplied against our P/O no. 8877. This order was placed after your representative left a sample pen with us in July, which we found to be very good quality. We have compared the performance of this sample with that of a number of pens from this batch, and there is little doubt that many of them are faulty – some of them leak and others blot when writing.

The complaints we have received relate only to pens from the batch mentioned. Customers who bought pens before these have always been pleased. We wish to return the unsold balance, amounting to 337 pieces, and to have them replaced with pens of better quality.

Please let us know what arrangements you wish us to make for the return of these unsuitable pens.

Regards,

Jack Turner
Purchasing Manager

我們最近收到了客戶對有關您提供的鋼筆的抱怨。這些筆顯然不令人滿意，因此在某些情況下，我們不得不退款給客戶。

這些鋼筆是我們訂單編號 8877 訂購的 500 支筆其中的一部分。當初我們下訂單是因為您的業務代表提供了一支樣品筆給我們，我們發現它的品質非常好。我們已經比較了這個樣本和該批次中的許多筆，發現當中許多筆有瑕疵，有些會在書寫時會漏水、有些會產生墨水汙漬。

我們收到的投訴僅涉及上述批次中的筆。在此之前購買鋼筆的顧客一直很滿意品質。我們希望退還未售出的部分，共 337 支，並換成品質較好的鋼筆。

請回信告知您希望我們如何退回這些不合適的筆。

範例 14-3　買方：抱怨品質不佳 II

To lindachen@tbi.com.tw
From jackturner@pioneer.com
Cc
Bcc
Subject Quality issue for PO#1332
Attachment

Dear Linda,

We have received PO#1332. We found the quality is far interior to the approval sample and the quantity has shortage of 10 cartons. In addition, 50 of the cartons were badly damaged due to fragile packing. Enclosed is the survey report for your reference, which shows damage resulting from faulty packing, shortage and defective quality. Under this awkward situation, we are threatened with the serious loss of our clients.

This is a very serious problem. Please take this matter into your careful consideration and give us an explanation as to how solve this problem. It seems that a reasonable compensation cannot be avoided as the damages were caused by errors made from your end. The charges incurred for the Survey Report and the repair or re-shipping of the goods will be at your expense.

We look forward to your reply soon.

Regards,

Jack Turner
Purchasing Manager

翻譯

我們收到了訂單 1332 的商品，但是我們發現商品的品質比確認的樣品差，並且數量短缺了 10 箱。除此之外，由於包裝不當，有 50 個外箱嚴重受損。附件是我們的調查報告供您參考，該報告顯示包裝不合格，缺貨和品質不佳。在此這種情形下，我們將有可能損失我們的客戶。

這是一個非常嚴重的問題。請仔細思考並告知貴公司如何解決這個問題。 因為貴公司所造成的問題，因此合理的賠償是不可避免的。賠償部分包含調查報告的費用以及貨物的維修和重新運輸商品的費用。

我們期待著您的回覆。

範例 14-4　買方：交運錯誤商品

To lindachen@tbi.com.tw
From jackturner@pioneer.com
Cc
Bcc
Subject Delivery of wrong goods
Attachment

Dear Linda,

We received our container yesterday and found two issues:

1. There are 20 cartons of P/N AA-1223, which we did not order. We assume that your factory has made a mistake in loading another's customer's goods into our container. Please advise what we should do with these goods.

2. P/N AA-1332, we ordered 100 cartons but just received 80 cartons. We need you ship the shortage by air freight at your expense immediately.

We are waiting for your prompt reply.

Regards,

Jack Turner
Purchasing Manager

我們昨天收到了我們的貨櫃，但發現兩個問題：

1. 裡面有 20 箱的商品，商品編號 AA-1223，這個商品我們沒有訂購。我們認為可能是您的工廠將另一位客戶的商品包裝到我們的貨櫃裡了。請告知我們應該如何處理這些商品。

2. 商品編號 AA-1332，我們訂購了 100 個箱，但只收到 80 箱。我們急需這些商品，因此您要立即空運給我們，但是運費要由您支付。

等待您立即的答覆。

範例 14-5　買方：產品不良要求退費或換貨

To	lindachen@tbi.com.tw
From	jackturner@pioneer.com
Cc	
Bcc	
Subject	Request refund or replacement
Attachment	

Dear Linda,

It is with regret that I inform you of some serious problems with your delivery of two cartons of leather sandals of PO 1122, received last week.

When I opened the consignment today, the sandals appeared to be moldy. Further inspection confirmed this, and though my staff and I tried to clean them, the sandals could not be restored to a salable condition. I can only assume that the low quality of the leather caused the mold growth.

I have always been happy with the quality of your goods, however, I am afraid I have to ask for a refund or replacement this time. Please advise how you will arrange this matter.

Regards,

Jack Turner
Purchasing Manager

我很遺憾地告訴您，上週收到的訂單 1122 的兩箱皮革製涼鞋有一些嚴重的問題。

今天我打開貨物時，涼鞋已發霉。我們做了進一步的檢查，更證實了這一點，雖然我的員工和我試圖清理它們，但是涼鞋還是無法恢復到可銷售的狀態。我只能假設可能是皮革的品質不佳而引起了黴菌的生長。

我一直對您的商品質量感到滿意，但是，這次恐怕我不得不要求退款或更換商品。請告知您如何處理此事。

範例 14-6　買方：抱怨未收到商品

To: lindachen@tbi.com.tw
From: jackturner@pioneer.com
Cc:
Bcc:
Subject: Shipment of PO AA1223
Attachment:

Dear Linda,

I placed an order (P/O no. AA 1223) on Sep. 25 and received confirmation from you on Sep. 30. We have still not received advice of delivery, so we are wondering if our order has been overlooked.

This delay is causing considerable inconvenience, so we hope you can complete the order immediately, otherwise we shall have no option to cancel it and obtain the goods from other suppliers.

Regards,

Jack Turner
Purchasing Manager

翻譯

我於 9 月 25 日下訂單（訂單編號 AA1223），並於 9 月 30 日收到您的確認。但是我們尚未收到您確認交期的通知，所以我們想知道您是否確實執行我們的訂單。

如果交貨延誤將造成我們相當大的不便，所以我們希望您能夠立即完成訂單。否則我們只好取消訂單，並從其他供應商那裡購買。

14.2 抱怨信的處理與回覆

在所有的商業信函中，最難處理的就是抱怨信 (complaint letter)，因為此類信函大都會牽扯到賠償或是損失。一般收到客戶的抱怨信函，供應商應該仔細聆聽，並妥善處理抱怨的事件，以免客戶流失或是轉向其他供應商。一般的觀念都認為「顧客永遠是對的」(Customer is always right.)，雖然實際上的情況不一定是如此，但是一開始面對客戶的抱怨時，我們至少必須假設「客戶也許是對的」(Customer may be right.)，特別是對於公司的大客戶來函抱怨時，更需特別小心謹慎處理。

對抱怨信的回覆有幾個一般的原則，但是實際上要如何回覆還是應該先檢視客戶的抱怨是否合理，需先判斷客戶是否有正當的理由提出這樣的抱怨。收到客戶的抱怨信時，第一件事就是要先確認信中所陳述的事件，以確定我方是否真有這樣的疏失，以及客戶的抱怨是否合理正當，之後再決定如何回應。

下表是客戶抱怨的項目與責任歸屬範圍：

抱怨項目	責任歸屬			
	賣方	買方	運輸公司	保險公司
貨品損壞	×		×	×
出錯貨品	×			
品質不良／瑕疵	×			
數量短缺	×			×
尺寸或是規格不符	×			
包裝不良	×		×	×
出貨延遲	×			
裝船延誤	×		×	
文件錯誤	×			
違反合約	×	×		
取消合約	×	×		
市場不景氣，客戶退貨		×		
卸貨卸錯港口			×	

另外國際條約亦有一般通則的規定，如買賣雙方對商品有爭議時：

1. 買方抱怨必須於貨到港口 21 日內提出，並附上證明文件 (Any claim by Buyer shall reach Seller within 21 days after arrival of goods at the destination stated in B/L accompanied with satisfactory evidence thereof.)

2. 但如有負責維修保養的保固期限的產品例外，在保固期間內，商品有任何問題，賣方仍需負責。
3. 買方使用不當造成的損失，賣方不負責賠償 (Seller shall not be responsible for damages that may result from the use of goods.)
4. 賣方最大的賠償金額不得超過不良品的價值 (Seller shall not be responsible for any amount in excess of the invoice value of the defective goods.)。除非訂單上買賣雙方另有約定賠償條款，則為例外情況。

1. 對合理的抱怨信的處理與回覆

如果確認客戶的抱怨是合理的，首先要先確認公司對於此類的抱怨有何既定的處理政策。依情況的不同，可以告知客戶我們所能處理的方式以及公司已經採取哪些步驟來處理此一問題；又或是公司可以採取不同的處理方式，但是要看客戶希望採取哪一種方式。因此回覆的信涵應具備下列的原則：

(1) 快速：儘快回覆，不要讓心有不買的客戶因等待而再增添不滿情緒。

(2) 謝謝：感謝客戶來函指出商品或是服務的問題。

(3) 行動：告知客戶你已經具體做了什麼或是採取了不同的方案供客戶選擇。讓客戶知道你關心他們的需求，也願意盡力滿足他們的需求。

(4) 道歉：在第一段就該為問題道歉，可以在信尾再做一次道歉以誠摯表達你的態度，這樣可以安撫客戶不滿的情緒。

通常供應商對客戶的抱怨信處理最多的項目是商品的品質問題，因此對客戶的抱怨可以：

1. 請客戶先寄回一個不良商品讓我方進行分析。
2. 判定出貨商品 (goods) 與確認的樣品 (approval sample) 品質是否一致，以釐清責任。
3. 如果是賣方責任，則賣方就應負賠償責任，通常處理的方式有三種方式：換貨 (replacement)、退款 (refund)、折扣 (discount)。而這些在客戶端的不良品也有幾種常見的處理方式，可以請客戶退回工廠、我方派人維修、轉賣當地其他客戶或是丟棄。這些處理方式端看商品價值而定。

範例 14-7　賣方：回覆抱怨數量短缺（Re 範例 14-1）

To	jackturner@pioneer.com
From	lindachen@tbi.com.tw
Cc	judywong@ttm.com.tw
Bcc	
Subject	RE:Quantity shortage for PO#1332
Attachment	

Dear Jack,

Please accept our apology for the shortage quantities of electric drills on PO#1332. We are going to send the replacement figures by air sometime next week. As soon as we have made the necessary arrangements, we will let you know the airline and other flight information.

We hope these goods will reach you in time to meet the anticipated demand for this article. If you have any further questions, don't hesitate to contact me.

Best regards,

Linda Chen
Sales Manager

翻譯

請接受我們對訂單號碼 1332 上電鑽數量短缺的道歉。我們將在下週安排空運出貨給您。一但我們安排好了，我們會立刻通知您航空公司和其他出貨資料。

我們希望您能及時收到這些商品，以滿足您對此商品的需求。如果您還有其他問題，請立刻與我聯繫。

範例 14-8　賣方：回覆抱怨抱怨品質不佳（接受抱怨）（Re 範例 14-2）

To	jackturner@pioneer.com
From	lindachen@tbi.com.tw
Cc	judywong@ttm.com.tw
Bcc	
Subject	RE:Quality issue for pens
Attachment	

Dear Jack,

Thanks you for your email regarding faults in the pens supplied to your order number 8877. We have been very concerned about this, and are glad that you brought this to our notice.

We have tested a number of pens from the production batch you mentioned, and agree that they are not perfect. The defects have been traced to a fault in one of the machines, which has now been rectified.

Please arrange to return to us your unsold balance of 377 pens. We will be happy to reimburse you for the cost of postage. We have already arranged for 400 pens to be sent to replace this unsold balance. The extra 23 pens are sent with our compliments. You will be able to provide free replacements of any further pens about which you may receive complaints.

Once again, our apologies for this inconvenience.

Best regards,

Linda Chen
Sales Manager

對於單價不高的商品，如果因為品質不佳而遭客戶抱怨，確認是我方的產品檢驗沒有檢測到，則寄出換貨商品時，可以多寄一些數量，這樣買方的感受會比較好。

翻譯

感謝您來函告知我們您的訂單號碼 8877 訂購的筆中有瑕疵。我們對這一件事非常關切，並高興您將此通知我們。

我們已經從您提到的生產批次中測試了許多筆，並且同意它們不夠完美。這是因為其中一台機器出現故障，目前已修理好了。

請將您未售出的 377 支筆退回給我們，我們會支付郵資。我們已經安排了 400 支筆更換，其中額外的 23 支筆免費給您，讓您免費換給提出抱怨的客戶。

再次為我們對此造成的不便深表歉意。

範例 14-9　賣方：回覆抱怨品質不佳（Re 範例 14-3）

To: jackturner@pioneer.com
From: lindachen@tbi.com.tw
Cc:
Bcc:
Subject: RE:Quality issue for PO#1332
Attachment:

Dear Jack,

We received with attention your mentioned issues regarding on PO#1332.
First, we sincerely apologize for the situation. However, we need to explain and state our arrangements as follows:

1. Quality issue: Concerning your claim the inferior quality, we have noticed the defects indicated on the Survey Report. We sincerely apologize for this situation and would like to make up our error by sending another lot next week. Is it acceptable for you, or would you prefer that we give you a 20% discount for these defective products?

2. Packing: We always use standard export carton which is not fragile as you mentioned, since we have exported our goods with this kind of carton to the other countries, and have not received any complaints or returns so far. As this is the first time we have heard of such an occurrence, we spent lots of time to figure out the cause of the damage. Our conclusion is that our goods were packed with the greatest care and we can only presume it was damaged in transit. You can claim it to the shipping company or insurance company.

3. Quantity shortage: We assure that the products were all shipped without any shortage from factory. We prepared the shipping documents according to the actual shipped items. If they were short on our side, the customs or shipping company would have found this when they counted the nos. and weight. This situation may have occurred during the transportation or transshipment, therefore, the shipping company or insurance company should take the responsibility.

We are sorry to put you in an embarrassing situation. For any errors we may have made, we will take full responsibility. However, as for the broken cartons and shortage, please make a claim to the shipping company and insurance company. If you need any assistance from our side, please inform us.

We hope you can accept our explanation and resolution, and then tell us of your comments or confirmation return.

Best regards,

Linda Chen
Sales Manager

翻譯

我們收到您提到有關於訂單號碼 1332 的問題。首先，我們誠懇地為這種情況發生感到抱歉。但是，我們必須解釋和說明這些情形的發生原因，以及我們如何處理的方式，以下是我們的說明：

1. 品質問題：關於品質不良的問題，我們注意到了調查報告中指出的缺陷。我們對這種情況表示誠摯的歉意，並會於下週寄給您一批商品以彌補我們的錯誤。請問這樣處理您可以接受嗎？還是您希望我們給您這些瑕疵品額外的 20% 的折扣？

2. 包裝：我們一直使用標準的出口紙箱，而且我們也用這種紙箱包裝運貨至其他國家，迄今為止還沒有收到任何投訴或退貨。由於這是我們第一次聽到有這種情況發生，我們花了很多時間找出損壞的原因。我們的結論是，我們的產品包裝得非常謹慎，我們只能假設它是在運輸過程中受損。您可以向運輸公司或保險公司索賠。

3. 數量不足：我們保證所有產品都在出廠時沒有任何短缺。我們根據實際裝運的物品準備了包裝明細表。如果商品在我們這裡有短缺，那麼海關或航運公司在計算這些箱數和重量時就會發現這一點。這種情況可能發生在運輸或轉運過程中，因此，船公司或保險公司應該承擔責任。

我們很抱歉讓您陷入如此的境地，對於我們可能犯的任何錯誤，我們將承擔全部責任。但是，對於破損的紙箱和短缺，請向運輸公司和保險公司提出索賠。如果您需要我方的任何幫助，請通知我們。

我們希望您能接受我們的解釋和解決方案，然後告訴我們您的意見以及確認。

範例 14-10　賣方：回覆交運錯誤商品（取回錯誤商品）（Re 範例 14-4）

To: jackturner@pioneer.com
From: lindachen@tbi.com.tw
Cc: judywong@ttm.com.tw
Bcc:
Subject: RE:Delivery of wrong goods
Attachment:

Dear Jack,

You mentioned there are 20 cartons shortage of P/O no. AA-1223 and the products were mixed up with our other customer's order. We apologize for the mistake that our factory loaded the products incorrectly.

We have arranged for the correct goods to be dispatched to you by vessel around April 10 at our expense. We could not make air shipment because the airfreight cost is extremely high. Please kindly understand. However, if you want to pay the airfreight cost, we can make shipment immediately. Let me know your decision.

Regarding the wrong products, please keep them and the contents until called for by our agents who have been informed of this situation.

We apologize for the inconvenience caused by our error.

Best regards,

Linda Chen
Sales Manager

翻譯

您提到訂單號碼 AA-1223 我們將其他客戶的商品混淆裝在您的出貨上，並且有 20 箱的短缺。對於我們的工廠發生裝運的錯誤，我們深表歉意。

我們已安排在 4 月 10 日左右將正確的貨物經海運寄給您，費用由我們承擔。由於空運成本非常高，因此我們無法以空運出貨，請諒解。但是如果您要支付空運費用，我們可以立即以空運送給您。請告知您的決定。

關於錯誤的商品，請先放在您那裡，我將請我們的代理人去處理。

對於我們的錯誤而造成您的不便，我們深表歉意。

範例 14-11　賣方：回覆交運錯誤商品（順道促銷錯運的商品）

To jackturner@pioneer.com
From lindachen@tbi.com.tw
Cc judywong@ttm.com.tw
Bcc
Subject Delivery of wrong goods
Attachment

Dear Jack,

Thank you for telling us about the mistake in our shipment to you last week. We sent our sincere apologies. I have checked with our shipping department and found they mislabeled your cartons which caused this mistake.

The 20 cartons of electric kettles you ordered are being shipped to you today. As far as the shipment of wrong product (blenders) we delivered to you in error is concerned, you can return them to us and we will pay the shipping charge. You may, however, wish to keep the blenders as well. They are a very popular item in your market this year.

Why not keep the blenders to see how market reacts? This item may increase your sales volume this year. Please let me know your decision.

Best regards,

Linda Chen
Sales Manager

對於瑕疵或是出錯的商品，若是商品價值不高，最好在當地解決或是想辦法出售，不要再運回來，因為不只手續繁瑣而且費用亦高。除非是高價商品才需對方將商品寄回。

翻譯

感謝您告訴我們上週我們運送給您的商品內有錯誤。在此我們致上誠摯的歉意。我有詢問我們的運輸部門，發現他們標記錯了紙箱而導致了這個錯誤。

我們今天已將您訂購的 20 箱電熱水壺寄給您。我們送錯的商品（攪拌機），請退還給我們，我們將支付運費。但是，說不定您想留下攪拌器，因為這個商品今年在您的市場上是非常受歡迎。

您為何不留下攪拌機來看看市場的反應如何？這個項目可能會增加您今年的銷量。請告知您的決定。

範例 14-12　賣方：回覆產品不良要求退費或換貨（同意）（Re 範例 14-5）

To jackturner@pioneer.com

From lindachen@tbi.com.tw

Cc

Bcc

Subject RE:Request refund or replacement

Attachment

Dear Jack,

We received with attention your email mentioned the poor quality of leather sandals of PO no. 1122.

We are very sorry that the leather has mold on it which may be caused by our quality inspector did not notice it this time. This situation has never happened before. Of course, all this is no excuse. We sincerely apologize for this defective products and will remit US$520 by wire transfer to your account. Or, you prefer we make replacements? If you prefer to receive replacements, we can deliver the goods next week.

Please let me know what you decide.

Best regards,

Linda Chen
Sales Manager

翻譯

我們注意到您的來函中提到了訂單號碼 1122 的涼鞋的皮料品質不佳的問題。

對皮革上有黴菌，我們感到非常抱歉。這可能是我們的品管檢驗員這次沒有注意到。這種情況以前從未發生過。當然，這一切都不能成為推託責任的藉口。我們對這些有瑕疵的產品表示誠摯的歉意，並將退款並電匯 520 美元給您。或是您想我們再寄一批貨給您來替換？如果您比較喜歡更換，我們可以在下週就寄商品給您。

請讓我知道您的決定。

範例 14-13　賣方：回覆抱怨未收到商品（Re 範例 14-6）

To	jackturner@pioneer.com
From	lindachen@tbi.com.tw
Cc	
Bcc	
Subject	RE:Shipment of PO AA1223
Attachment	

Dear Jack,

Thank you for your email and I quite understand your annoyance at not having received PO AA1223.

I have immediately checked into this matter and am pleased to inform you that your goods are ready to ship. I will arrange the shipment by vessel next week and will send you the shipping information when available.

Once again, I am sorry not inform you the shipping information earlier and hope you are pleased with the goods when they are received.

Best regards,

Linda Chen
Sales Manager

翻譯

感謝您來函，我非常理解您沒有收到訂單 AA1223 的商品的困擾。

我已經立即去了解此事，並很高興地通知您，您的貨物已準備好出貨。我們將在下週安排裝運貨物出貨，並在出貨細節資料得到時，寫信告知您出貨明細。

再一次地抱歉沒有及時通知您送貨的訊息，並希望您在收到貨物時感到滿意。

2. 對不合理的抱怨信的處理與回覆

有時候很不幸的是客戶還是會提出不合理的抱怨。這時候我們可以體諒並尊重客戶的觀點，在撰寫回覆信時依舊要遵守之前說明的回覆原則。但是因為這是不合理的抱怨，因此這時候我們不用說明我方會採任何行動以解決此一問題，而是要解釋為什麼我方無法依客戶要求的方式進行補償。在這一類的回信中，謹慎周到非常重要。

客戶的抱怨不合理，並不代表我們就不應該提供任何協助。畢竟來函抱怨的客戶還是客戶，提供客戶完善的服務最終對公司還是有益的。

範例 14-14　賣方：回覆抱怨抱怨品質不佳（不接受抱怨）（Re 範例 14-2）

To: jackturner@pioneer.com
From: lindachen@tbi.com.tw
Cc: judywong@ttm.com.tw
Bcc:
Subject: RE:Quality issue for pens
Attachment:

Dear Jack,

We are sorry to learn from your email that you have had difficulties with the pens supplied to your order number 8877.

All our pens are manufactured to be identical in design and performance, and we cannot understand why some of them should have given trouble to your customers. It is normal practice for each pen to be individually examined by our Inspection Department before being passed into store. However, from what you said, it would seem that a number of the pens included in the latest batch escaped the usual examination.

While we certainly understand your concern, we cannot accept your suggestion to take back all the unsold stock from this batch. Indeed, there should be no need for this since it is unlikely that there are a large number of fault pens. We will gladly replace any pen found to be unsatisfactory, and on this particular batch we are prepared to allow you a special discount of 5% to compensate for your inconvenience.

I hope you will accept this as being a fair and reasonable solution of this matter. Please give me a call on 886-2-24587962 or send me an email if you have any further questions.

Best regards,

Linda Chen
Sales Manager

對於客戶的抱怨，並非照單全收。依客戶所提的求償方式，須根據公司的政策，而有所應對方式。但是就算是情況顯示不必接受客戶的抱怨，也應該要婉轉的表達，應該要表現出你了解客戶的立場，並小心解釋不得不拒絕的原因或是其他替代方案。

翻譯

我們很抱歉得知您收到的訂單號 8877 的筆有瑕疵。

我們生產的鋼筆在設計和功能都是相同的，因此我們不明白為什麼有些鋼筆會給您的客戶帶來問題。每一支筆在被送進庫房之前都經過品管部門一一檢查。但根據您的說法，最近這一批似乎沒有被檢查到。

我們確實了解您的問題，但我們不能接受您的建議，將所有未售出的筆回收。事實上，您並不需要如此做，因為有瑕疵的筆可能只是少數。我們很樂意更換瑕疵的筆，並對這些筆提供額外的 5% 特別折扣，以彌補您的不便。

希望您會接受此一公平且合理的解決方案。如果您還有其他問題，請打 886-2-24587962 與我聯絡。

範例 14-15　賣方：回覆產品不良要求退費或換貨（不同意）（Re 範例 14-5）

To	jackturner@pioneer.com
From	lindachen@tbi.com.tw
Cc	
Bcc	
Subject	RE:Request refund or replacement
Attachment	

Dear Jack,

Thank you for your email explaining the problem with your P/O no. 1122. We are very sorry that you are not able to use the products and we understand how frustrating it is to receive damaged goods.

However, in light of the fact that the consignments have been in your possession for the past 8 months, I am afraid we are unable to take responsibility for their current condition. Our quality guarantee covers our products while they are in our or our shipper's control (as shown on contract). Unfortunately, we have no way of knowing the condition of the sandals when they arrived. Since we have not received any other complaints about the quality of these particular sandals, we must entertain the possibility that your storage conditions contributed to the mold.

We do appreciate the problems that unsalable merchandise poses. Although we are unable to offer you a refund on your order, we would be happy to take a 5% discount off your next order and hope this unavoidable accident won't cause any unfavorable influence for your further orders.

Best regards,

Linda Chen
Sales Manager

翻譯

感謝您來信說明您收到的訂單號碼 1122 的問題。對於您無法出售這些涼鞋，我們感到非常抱歉，並且也了解收到瑕疵的商品有多麼令人挫折。

但是，由於這些商品在過去 8 個月以來一直在貴公司手上，恐怕我們無法對目前的商品情況負責。我們的品量保證範圍是包含商品還在我們的貨運上（如合同所示）。但是我們並不知道涼鞋到您手上時的狀況。由於我們沒有收到任何其他有關這些涼鞋品質不良的投訴，我們認為有可能是您的儲存條件不良而造成發霉的現象。

我們能夠理解涼鞋無法出售對貴公司帶來的問題。雖然我們無法為您提供訂單退款，但我們可以給您的下一個訂單提供 5% 的折扣。並希望這一事件不會對您未來訂單造成任何的影響。

常用句

買方

1. We have received serious complaints from clients in regard to your TV sets.
2. We regret having to inform you that we have had several complaints from customers concerning the poor material.
3. The goods received are much inferior to the sample in quality.
4. For goods of such inferior quality, we must claim a large allowance.
5. We are disappointed to inform you that we have received many complaints from our customers due to your defective products, which are far below our standard.
6. The goods we ordered on January 3rd have not yet been delivered.
7. We have to complain that the quality of the men’s shirts is worse than that of the sample that you sent last quarter.
8. I hope to receive a complete refund soon.
9. I hope you will investigate this and take the necessary action.

10. Your prompt settlement to this matter will be appreciated.
11. I look forward to hearing from you about what you can do to permanently stop this from happening in the future.
12. How do you propose to remedy this situation?
13. Your quick attention to the mentioned problems would be appreciated.
14. Please give this matter in your urgent attention.

賣 方

合理抱怨的回覆：

1. We are sorry to learn that this shipment has not given you full satisfaction.
2. I am sorry to hear that the goods sent under this order did not reach you until June 25.
3. Thank you for your message and we are sorry for the unfortunate mistake that was made.
4. We appreciate the opportunity to clarity this issue.
5. I am sorry about the distress this caused you.
6. In the circumstances, it is important that we make amends for your inconvenience.
7. Please accept our apologies for the delay and the inconvenience it has caused you.
8. If you keep the damaged goods, we could offer you 20% discount.
9. We will make a special allowance of US$20 per carton.
10. We will send replacements right away.
11. We will ship your replacement immediately.
12. We hope you will be satisfied with the arrangements we have made.
13. Once again, we apologize for the inconvenience.
14. We hope these arrangements are satisfactory and look forward to receiving your future orders.
15. Thank you once again for taking the time to write to us.

回拒不合理的抱怨：

1. I am sorry, but according to our policy, we cannot offer you a replacement.
2. We made a careful examination of our original samples and found no any evidence of inferior quality. We assure that the materials we used were of the highest quality.
3. Please kindly note and understand that the goods were shipped in good condition from our factory. It is impossible to have a shortage because they were weighed at the customs. You may refer and compare the weight shown on the clean B/L.
4. Because the problem occurred somewhere outside our area of responsibility, we are not able to refund the money to you.
5. Your claim has been passed on to our insurance company, who will get in touch with you soon.
6. As soon as we received your email, we got in touch with the shipping agents and asked them to look into the matter.
7. As this is due to force majeure, we suggest that you lodge your claim with the insurance company.
8. Since the damage is not our fault, we recommend that you file a claim with the insurance company.

練習題

1. 將下列的句子翻成英文

A. 關於您所供應的電視機架，我們已經收到很多嚴重的抗議。

B. 在此我們向您抗議所收到的貨物比樣品的品質差很多。

C. 很遺憾地通知您，關於您的貨物，我們收到許多抱怨，而且大部分已遭退貨。

D. 我們想得知您對商品品質不良的解釋，並想了解對本次事件您將如何處理。

E. 您的貨物比樣品差多了，而且顏色也略有不同。從所附檢驗報告中，您會承認這些貨物在品質上比我們要求的標準低。

F. 為了證明品質不良，我們附上羅德檢驗所 (Lloyd's) 的一份檢驗報告。您將會很快同意我們索賠的合理性。

G. 由於貨物品質過於低劣，我們必須要求賠償。

H. 為了證明我們的說法，附上一塊樣品及一塊昨天剛收到的質料以供比較。

I. 我們剛收到文件及貨物，很遺憾發現少了 2 箱貨。

J. 核對收到的商品時，發現發票中的許多樣商品均沒有放入。

K. 收到的貨數與發票不符，短少 5 箱。

L. 很失望的發現您提供的商品與樣品不符。

M. 貨物與樣品不符。

N. 令人驚訝的是，有些貨物已經壞掉了。

O. 貨到時，已有 2 箱受損，我們也據此在通知單上註明。

P. 我們 7 月 15 日所訂的商品，收到時損壞情形很嚴重。

Q. 在打開的箱子中，發現半數的盤子都已打破了。

R. 由於包裝不良，有些貨物已嚴重損壞，因而不得已以極低的價格拋售。

S. 出現的貨品損壞是因為包裝錯誤。

T. 我們 4 月 19 日訂的 2000 支手機都沒出貨，請處理。

U. 交期延誤給我們帶來極大的不便，因為我們答應客戶早點出貨。

V. 自得到您的出貨通知至今已逾一個月，但是至今我們仍未收到船公司的任何通知。

W. 貨物比原先的行程慢了 7 天才進港。

X. 本公司要求貴公司做出相對於本公司損失之賠償。

Y. 請通知貴公司是要取回貨物，或是要本公司以對折方式予以銷售。

Z. 本公司已準備留下這些不良品，但須大幅降價才行。

AA.由於貨物品質低劣，即使貴公司於價格上作出讓步，本公司客戶也拒絕接受這些商品。

AB.本公司要將此項索賠提出予以仲裁。

AC.如果商品幾日內仍找不到的話，我們只好提出索賠處理了。

AD.請調查此事並儘速處理。

AE.您的立即回覆，將不勝感激。

2. 將下列中文書信翻譯成英文

我們很遺憾的通知您，貴公司於 2 月 3 日送達的 10 箱涼鞋有嚴重的品質問題，鞋面都發霉了。

這些商品的包裝看起來完好無缺，我沒有質疑就收貨，開封後才發現上述損壞的商品，我猜應該是在包裝前的某一個階段處理不慎所造成的。

隨函附上損壞品清單，希望您能儘快予以更換。目前這一批商品暫時存放在我這裡，請告知要如何處理。

3. 請寫一封抱怨信，表示收到訂單號碼 PO-8541 的出貨，但是發現商品的尺寸與顏色不對，而且有些箱子有破損的情形，請儘快告知處理方法。

4. 請就第 3 題回覆，請客戶寄回不良樣品，以便分析原因，避免雙方誤解，再告知解決的方法。

5. 請就第 3 題回覆，抱歉品質與包裝不良，願給 2% 折扣。

6. 請寫一封回函告知客戶，對於客戶抱怨的包裝與數量短缺問題，我們正在調查，將於 3 天內回覆。

CHAPTER

15 代理

Agency

在現代的國際商務中，透過代理權的取得去銷售一項暢銷的商品是一種常見的手法，藉以增進公司的銷售額。另一方面，如果我們可以透過國外的代理商銷售商品，不僅可以降低先期在海外設置分公司的成本，亦可增加公司在海外的銷售總額。

一般代理的模式可以分為四種：

1. **銷售代理 (Sale Agent)**：在某些國家或是地區，因賣方不熟悉當地的法令、文化或是生活習慣，為了在當地推銷及販售商品，會在當地尋找一個可靠的代理商，協助銷售與推廣商品。代理的商品可以專指一項也可以多項，端看雙方的代理約定。一旦雙方簽訂合約後，賣方即不可再銷售給該簽約中所約定的國家或是區域，代理商也不可再向其他供應相同商品的廠商購買。
2. **採購代理 (Purchase Agent)**：有些進口商在某些區域會購買許多商品，如果商品種類與數量繁多，每一項商品都要進口商一一去搜尋、比價則耗時費力又花費高。因此，為確保貨源穩定與更強而有力的搜尋能力及採購的順利，進口商就會在當地找一可靠的採購代理，協助其詢價、索樣、下單與驗貨等相關業務。
3. **進口銷售代理 (Import Exclusive Agent)**：某些國外廠商製造的商品欲打入某個國內市場，或是進口商覺得某些產品在國內有極佳的銷售潛力，為了避免競爭以及獲取較佳的利潤，就會去向國外的廠商爭取獨家代理權 (Exclusive Agent or Sole Agent) 或是總經銷權 (Distributor)。
4. **佣金代理 (Commission Agent)**：是以佣金方式作為買賣交易基礎的一種方式。

雙方確認代理的關係之後，則會簽定代理合約，通常代理合約的內容需含下列的基本條款：

1. 甲、乙雙方資料 (Party A, Party B)：公司全名及地址。
2. 見證人 (Witness)：可為律師、第三者或是甲乙雙方，一般基於互信原則，都以甲乙雙方互為見證。
3. 生效日 (Effective Date)：合約起始日，大多以雙方簽字後即生效，亦可於合約上註明生效日。
4. 合約期效及年限：雙方可自行約定，通常以一年期為多。到期後，雙方再談續約與否，若欲中途解約，則需雙方同意才生效。

5. 終止日 (Termination)：如為長期合約（超過一年以上），多半會註明到期日，以防萬一中途有一方要取消合約。終止日的算法是以合約中任何一方以掛號書面提出後，60~90 天後此合約才終止。
6. 代理的產品 (Exclusive Products)：產品的名稱或型號，或是產品大類名稱。
7. 代理的區域 (Exclusive Area)：指代理商的國家或是其中某個區域。若是代理某特定地區時（如亞太區域），則需列出代理區域的國家名稱。
8. 最低銷售額 (Minimum Sales Amount)：成為代理商則需保證購買量或是購買金額。雙方會依市場的潛力及購買力，約定金額或數量。
9. 付款方式 (Payment)：如信用狀、T/T 出貨前或是出貨後，雙方達成協議，並詳述於合約中。
10. 抱怨 (Claim)：合約上有時會註明有糾紛時要如何解決。如果沒有特別註明，就按國際條約執行。
11. 仲裁 (Arbitration)：在每個國家都有國際仲裁協會 (International Arbitration Association)，附設於商會或是工會中，當事人雙方如有無法解決的糾紛發生時，可請任何一國的仲裁協會來裁定。但因此法費時又花錢，通常雙方都會私下解決。
12. 簽字 (Signature)：合約上一定要有雙方的簽字才有效。

範例 15-1 請求代理

To robertellis@pioneer.com
From johnwang@tbi.com.tw
Cc
Bcc
Subject Sole Agent
Attachment

Dear Mr. Ellis,

We have noticed that you have no agent in Taiwan, and we would like to offer you our services.

We have been selling a variety of durable goods to wholesalers and large retailers in Taiwan for past decades and have built up a considerable number of well-established connections. We believer, we have the ability to expand our sales, and if you agree to grant us exclusive rights to distributorship, we will devote full attention to establishing your products in this market.

If you are interested in our proposal, we would be more than glad to provide our references. We are looking forward to your reply.

Sincerely,

John Wang
General Manager

翻譯

我們有發現貴公司在台灣還沒有代理商，我們想為您提供這一項的服務。

我們在台灣經銷耐久材給大型經銷商與零售商已超過幾十年了，我們已建立了相當多的良好關係。我們堅信，我們有能力擴大我們的銷售，如果您同意授予我們獨家代理權，我們將盡全力在這個市場銷售您的產品。

如果您對我們的提議感興趣，我們將非常樂意提供我們的公司資料。期待您的回覆。

範例 15-2　請求獨家代理

Taiwan Bear International
3F, No. 11, Park Avenue II
Science-Based Industrial Park
Hsin-Chu 30075, Taiwan
Tel: 886-03-5798888
Fax: 886-035978891
www.tbi.com.tw

February 19, 20xx

Pioneer Corporation
3080 Bowers Avenue
Santa Clara, CA 95054
USA

Dear Mr. Sir/Madam,

Subject: EXCLUSIVE AGENT

We are currently expanding our retailer store, and want to extend our product range.

We are particularly interested in your range of regular and mountain bikes, as well as foldable bikes. We would like to receive your trade catalogue and terms of sale and payment. I believe your products are not yet offered by any other dealers in our country, and if we decide to introduce them we should like sole distribution rights in this area.

I look forward to hearing from you soon.

Sincerely yours,

John Wang

John Wang
General Manager
Email:johnwang@tbi.com.tw

翻譯

我們目前正在擴大我們的零售店，並希望擴大我們的產品範圍。

我們有特別注意到您的商品，包含標準款的自行車、越野自行車和折疊式的自行車。我們想請您寄給我們您的商品型錄以及銷售和付款條件。 我相信您的產品在我國還沒有透過任何的經銷商銷售，如果我們決定銷售您的商品，我們希望可以取得貴公司在本國區域的獨家經銷權。

期待早日收到您的消息。

範例 15-3　回覆請求代理（接受）（Re 範例 15-1）

To	johnwang@tbi.com.tw
From	robertellis@pioneer.com
Cc	
Bcc	
Subject	Re:Sole Agent
Attachment	

Dear Mr. Wang

Thank you for your email in requesting to be our sole agent in Taiwan. In view of the steady increase in the demand for our cell phone, we have decided to appoint you as an agent to handle our product in your country.

We have no doubt that there is a vast potential market for exploration, and that a really active agent can develop a profitable business in this line. Attached is our contract for sole agent in a specific country. Please review terms and conditions and advise if you agree the contract.

If you have no questions about the agent contract, after we both sign the contract, you will be our sole agent in Taiwan. We would appreciate an early reply as we hope to reach a quick decision.

Sincerely,

Robert Ellis
Sales Director

翻譯

感謝您來函請求成為我們在台灣的獨家代商。鑑於我們手機的需求穩步增長，我們決定授予您成為我們在貴國的代理商以銷售我們的產品。

我們相信在貴國潛在著極大的商機，而且一個非常有活力的代理商可以在這一市場上發展相當的銷售量。附件是我們在特定國家的獨家代理合約。請您看一下合約內容，並告知您是否同意此合約。

如果您對代理合約沒有任何問題，我們雙方簽訂合約後，您將成為我們在台灣的獨家代理商。我們希望能早日得到您的回覆，因為我們希望能夠儘快做出決定。

範例 15-4　回覆請求代理（拒絕）（Re 範例 15-1）

To	johnwang@tbi.com.tw
From	robertellis@pioneer.com
Cc	
Bcc	
Subject	Re:Sole Agent
Attachment	

Dear Mr. Wang

We thank you for your email in requesting to be a sole agent for our products in Taiwan. As part of our efforts to keep down manufacturing costs, I am sure you will understand that we must increase sales by distributing through as many distributors as possible. Distributors in other areas appear to be very satisfied with their sales under this arrangement, it appears to be working very well.

Therefore, our company does not consider sole agent in your country right now. However, we will welcome your orders if you are interested to sell our products with other distributors in your market. I hope we can look forward to receiving your orders soon, and would be glad to include your name in our list of approved distributors.

Sincerely,

Robert Ellis
Sales Director

翻譯

感謝您來函請求成為我們在台灣的獨家代理商。相信您一定了解，為了努力降低我們的成本，我們只能透過許多的代理商銷售我們的商品來增加銷售額。因為我們其他區域的經銷商對於這樣的銷售模式非常滿意，而且銷售狀況也非常好。

因此我們公司目前沒有考慮在貴國設置獨家代理商。但是如果您對銷售我們的商品有興趣，我們非常歡迎您下訂單給我們，並跟在貴國的其他經銷商一起銷售我們的商品。我希望能儘快收到您的訂單，並很樂意將您的名字列入我們合作的經銷商名單。

範例 15-5　回覆請求代理（拒絕：當地已有代理商）（Re 範例 15-1）

To	johnwang@tbi.com.tw
From	robertellis@pioneer.com
Cc	
Bcc	
Subject	Re:Sole Agent
Attachment	

Dear Mr. Wang

Thank you for your interest in being our exclusive agent in Taiwan.

We appreciate your proposal, however we have to inform you that we already have an exclusive agent in your country. We are forced to decline your proposal in compliance with our policy of setting a sole agent by “One Agent, One Area.”

Thank you for your understanding.

Sincerely,

Robert Ellis
Sales Director

翻譯

感謝您有興趣成為我們在台灣的獨家代理。

我們感謝您的提議，但我們必須告知您，我們在您的國家已經有獨家代理商了。因此我們無法接收您的建議，因為我們公司的政策是「一個區域只能有一個代理商」。

感謝您的理解。

範例 15-6　回覆請求代理（拒絕：合作一段時間後再談）（Re 範例 15-2）

Pioneer Corporation
3080 Bowers Avenue
Santa Clara, CA 95054, USA
Tel: (408) 576-8888
web:www.pioneer.com

March 19, 20XX

John Wang
Office of the President
Taiwan Bear International
3F, No. 11, Park Avenue II
Science-Based Industrial Park
Hsin-Chu 30075, Taiwan

Dear Mr. Wang:

Subject: Exclusive Agent

We have received your letter of Feb. 19 in which you have indicated that you are interested in being our exclusive agent in Taiwan.

We thank you for your proposal, but we would prefer to discuss this after two years so that we may understand each other more and have actual business contact first. I will suggest that you deal with us on a case-by-case basis first, until we get more familiar to each other.

In order to establish a favorable business relationship with you, we will do everything we can to give you our best offer. We hope our products may meet with success in your market, then, after having several regular orders, we may discuss the sole agent issue further.

Thank you for your understanding and hope you will agree with our opinions. We look forward to your further comments or order confirmation soon.

Sincerely yours,

Jack Turner

Jack Turner
Purchasing Manager

我們收到了您 2 月 19 日的來信，表示您有興趣成為我們在台灣的獨家代理商。

我們感謝您的提議，但我們能先跟您建立商業關係，讓彼此有更多的了解之後，兩年後再討論這個問題。我建議您可以先以個案的方式與我們先接洽，讓我們雙方更了解彼此。

為了與您建立良好的業務關係，我們將盡我們所能提供給您最好的報價。希望我們的產品能夠在您的市場上有很好的銷售量，這樣您再下了經常性的訂單後，我們就可以進一步討論獨家代理的問題了。

感謝您的理解，並希望您能同意我們的觀點。期待您的意見與訂單。

範例 1：代理合約書

DISTRIBUTION AGREEMENT

AGREEMENT made as of the _____ day of _______________, 20____, by and between *NAME OF SUPPLIER*, having its business address at ___________________ ____________________________________ (hereinafter referred to as "SUPPLIER") and *NAME OF DISTRIBUTOR*, having its principal place of business at ________________ _____________________________ (hereinafter referred to as "DISTRIBUTOR").

WITNESSETH

WHEREAS

A. SUPPLIER is the producer of certain wines from the state of *STATE* as more particularly identified on the price list attached hereto, and made a part hereof, as schedule A (hereinafter referred to as the "Products");

B. DISTRIBUTOR desires to secure from SUPPLIER, and SUPPLIER is willing to grant to DISTRIBUTOR, the exclusive right to sell and distribute SUPPLIER'S Products in the United States of America with the exception of the state of *STATE* and direct retail, airline or consumer sales and sales exported out of the country from SUPPLIER'S *STATE* winery (hereinafter referred to as the "Territory").

NOW THEREFORE, it is mutually agreed as follows:

1. **SUPPLIER** hereby appoints DISTRIBUTOR as its sole and exclusive distributor for the term of this Agreement for the sale and distribution of the Products in and throughout the Territory. DISTRIBUTOR will maintain, or cause to be maintained, a sales staff for the distribution of products handled by DISTRIBUTOR, including the Products, and DISTRIBUTOR shall use its best efforts to promote the sale and distribution of SUPPLIER'S Products.

2. **SUPPLIER** will not ship the Products, or any other wines bearing the same or similar trademark, signature or identification anywhere on the package, to the Territory except under the order or by the direction of DISTRIBUTOR. It will refer to DISTRIBUTOR any and all orders or inquiries for the Products that it may receive for shipment to the Territory, or orders which are intended for eventual shipment to the Territory.

3. **SUPPLIER** will fill promptly and to the best of its ability all orders for the Products received from DISTRIBUTOR. The price to DISTRIBUTOR shall be based on delivery to DISTRIBUTOR'S warehouse and shall include a mutually negotiated delivered price to said warehouse. SUPPLIER and DISTRIBUTOR shall negotiate any price increases for the Products at least 60 days prior to the effective date of any such increase. DISTRIBUTOR shall have the right to order one months supply of the Products at the current price prior to any increase. Payment in U.S. dollars shall be made by DISTRIBUTOR 90 days from the date of delivery to DISTRIBUTOR'S warehouse.

4. **DISTRIBUTOR and SUPPLIER** shall agree on an annual basis, or more frequently if required, as to the prices at which DISTRIBUTOR shall sell the Products to its customers. SUPPLIER will furnish to DISTRIBUTOR, promptly upon request, any and all authorizations that may be required by any governmental authority in connection with the sale and distribution of the Products in the Territory, provided that SUPPLIER is responsible for obtaining or maintaining said authorizations.

5. **Pursuant** to paragraphs 3 and 4 hereof, SUPPLIER and DISTRIBUTOR shall agree on SUPPLIER'S price to DISTRIBUTOR and DISTRIBUTOR'S price to its customers. In the event that SUPPLIER and DISTRIBUTOR cannot agree on either price within 30 days of commencement of the negotiations, the prices then in effect for each of said prices will be increased by an amount equal to the change in the Consumer Price Index-All US over a period of months equal to the number of months since the last price increase for each price.

6. SUPPLIER warrants, represents and agrees that all shipments of the Products sold or shipped under this Agreement shall be of first quality, suitable for beverage consumption,

properly bottled and packaged in *STATE*, free from foreign matter, whether or not prejudicial to health, and will be bottled and packaged in conformity with applicable laws, regulations and requirements in effect within the Territory.

7. SUPPLIER will, upon demand, promptly execute such documents and perform such acts as may be necessary so as to prevent any products labeled in imitation or simulation of the Products from being distributed in the Territory.

8. The term of this Agreement shall be for a period of two years commencing on *DATE*, and terminating on *DATE*, and shall thereafter continue in effect unless either party shall notify the other of its intention to terminate this Agreement by giving at least 12 months written notice prior to any specified termination date. Either party shall have the option to terminate this Agreement after six months of the notice period by paying to the other party a sum equal to one-half of the case volume of the previous calendar year multiplied by $*DOLLAR* per case. However, in the event of a breach of any of the terms and provisions of this Agreement, either party may terminate this Agreement by giving the other party 90 days written notice provided said notice shall set forth the breach being claimed as the basis for termination. If the offending party cures the breach being claimed within said 90-day period, the notice of termination shall be void and this Agreement shall continue in full and force and effect.

9. Notwithstanding the provisions of paragraph 7 hereof, SUPPLIER shall have the right to terminate this Agreement upon 60 days written notice in the event that DISTRIBUTOR shall:

a. be declared bankrupt or enter a voluntary petition for bankruptcy or in any way enter into a compromise or agreement for the benefit of its creditors;

b. fail to meet at least 90 percent of the mutually agreed upon sales performance goals set forth in Schedule B, attached hereto and made a part hereof;

c. fail to maintain in good standing all Federal and state licenses and permits necessary for the proper conduct of its business;

d. change or in any way be affected by a change in the majority ownership of its business.

10. DISTRIBUTOR, upon request from SUPPLIER, will furnish SUPPLIER with available sales and depletion reports and details of all promotional and sampling programs with respect to the Products. DISTRIBUTOR will discuss with SUPPLIER any proposed changes in its distributor network at least 30 days prior to any such change.

11. Upon termination of this Agreement by either party, SUPPLIER shall repurchase, or cause its successor representative to purchase, as of said termination date, DISTRIBUTOR'S then existing inventory of SUPPLIER'S Products at DISTRIBUTOR'S laid-in cost, provided DISTRIBUTOR has properly stored and maintained the inventory of the Products in a saleable condition.

12. This Agreement is the entire agreement between the parties, cannot be changed orally, and neither party has made any representations or promises to the other which are not expressed in this Agreement.

13. No waiver of a breach of the terms of this Agreement shall be effective unless made in writing, and no such waiver shall be deemed a waiver of any other existing or subsequent breach. No modification of this Agreement shall be of any effect unless set forth in writing.

14. All the provisions of this Agreement are made subject to all applicable laws, regulations, rules or requirements of the Government of the United States of America or agencies of said Government, and in the performance of this Agreement, each of the parties hereto agrees to comply therewith.

15. All notices shall be sent prepaid either by mail or facsimile addressed to the respective parties at the address hereinabove set forth, unless they shall otherwise notify in writing.

16. This Agreement is an *STATE* contract and shall be governed by and construed in accordance with the laws of the state of *STATE*. Any controversy or claim arising out of or relating to this Agreement or the breach thereof shall be settled by arbitration in

STATE in accordance with the rules of the American Arbitration Association then in effect, and judgment upon the award rendered by the arbitrator or arbitrators shall be final and binding upon the parties hereto.

17. If arbitration is required to enforce or to interpret a provision of this Agreement, or otherwise arises with respect to the subject matter of this Agreement, the prevailing party shall be entitled, in addition to, other rights and remedies that it may have, to reimbursement for its expenses incurred with respect to that action, including court costs and reasonable attorneys' fees at trial, on appeal;, and in connection with any petition for review.

18. This Agreement shall not be assigned by either party hereto.

IN WITNESS WHEREOF, the parties hereto have caused this Agreement to be executed as of the day and year first above written.

SUPPLIER	DISTRIBUTOR
By:_______________	By:_______________
Title:_______________	Title:_______________

範例 2：代理合約書

AGENT AGREEMENT

THIS EXCLUSIVE DISTRIBUTION AGREEMENT (this "Agreement" or this "Exclusive Distribution Agreement) is entered into effective as of [DATE] (the "Effective Date") by and between [SUPPLIER.Company] ("Supplier") and [DISTRIBUTOR.Company] ("Distributor").

The parties agree as follows:

1. APPOINTMENT

(1) Subject to the terms and conditions of this Exclusive Distribution Agreement, Supplier appoints Distributor, and Distributor accepts such appointment and agrees to act as Supplier's exclusive distributor of the Supplier Products (defined below) within the geographical territory defined as follows (the "Territory"):

(2) Distributor agrees to exercise its best efforts to (a) promote the sale of and obtain orders for the Supplier Products in the Territory; (b) abide by Supplier's policies and procedures with regard to the purchase, sale and support of Supplier Products; and (c) conduct its business in a manner that reflects favorably at all times on the Supplier Products and the good name, goodwill and reputation of Supplier or its affiliates. Distributor acknowledges and agrees that it has no rights or claims of any type to the Supplier Products, or any aspect thereof, except such rights as are created by this Exclusive Distribution Agreement. Distributor agrees that it shall not and is not authorized to promote, resell, deliver, install, service or therwise support the Supplier Products outside of the Territory.

2. PRODUCTS AND PRICING

Supplier Products consist of the items or classifications of items listed in this Section below, and the purchase price or license fee to Distributor of all Supplier Products delivered pursuant to this Agreement shall be as set forth in this Section below.

3. APPROVALS

Distributor shall obtain, at its own expense, such approvals, consents, certifications, permits and other authorizations, including all approvals as are required to qualify the

Supplier Products for sale and use in the Territory for all purposes, both governmental and non-governmental (collectively, the "Approvals"), as soon as is reasonably practicable; provided, however, that Supplier shall not be obligated to deliver any Supplier Products unless and until Distributor provides Supplier with satisfactory evidence that such Approvals have been obtained. Supplier agrees to cooperate with Distributor to obtain such Approvals.

4. EXCLUSIVITY

Supplier's appointment of Distributor in Section 1 of this Agreement is an exclusive appointment to distribute the Products in the Territory. Supplier shall not independently advertise, solicit and make sales of Supplier Products, support Supplier Products or appoint additional distributors for Supplier Products in the Territory.

5. SALES TARGETS

Distributor shall use reasonable commercial efforts to purchase and sell during each calendar year at least the dollar value of Supplier Products listed in this Section below (the "Annual Target"). If Distributor does not purchase and sell the Annual Target during a given calendar year, Supplier may terminate this Agreement effective immediately upon notice to Distributor (but Supplier may not require Distributor to purchase or sell any additional Supplier Products in order to meet the Annual Target Amount).

6. SALES OUTSIDE OF THE TERRITORY

Distributor shall promote the sale of Supplier Products in the Territory on its website. Notwithstanding the foregoing sentence, Distributor shall not actively advertise or actively solicit orders for Supplier Products outside of the Territory. In the event Distributor receives an order from outside its Territory, Distributor will work with Supplier to fulfill the order in a manner financially beneficial to Supplier, Distributor and the distributor located in the region where the order originated (as determined by Supplier in its sole discretion and in compliance with applicable law).

7. NO THIRD PARTY AGENTS

Distributor shall not sell/license the Supplier Products through third parties (such as original equipment manufacturers, distributors, value added resellers or other dealers or agents) without Supplier's prior written consent to the proposed relationship (including the specific terms of such relationship).

8. ORDERS

All orders will be transmitted by Distributor to Supplier and shall be subject to acceptance in writing by Supplier. Supplier may (in its sole discretion) refuse acceptance of any order. Each order submitted shall constitute an offer by Distributor to purchase or license the Supplier Products described in such order and, upon acceptance by Supplier, shall give rise to a contractual obligation of Distributor to purchase or license the said products on the terms and conditions set forth in this Agreement. Conflicting, inconsistent or additional terms or conditions contained in any order submitted by Distributor shall not be binding unless Supplier specifically accepts such terms or conditions in writing. All expenses arising out of the change or cancellation of an order after acceptance by Supplier, including the cost of diversion, cancellation or reconsignment of shipments, and any reasonable restocking charge, shall be paid by Distributor to Supplier, on demand.

9. PAYMENT AND DELIVERY

The purchase price shall be quoted and payable in U. S. Dollars to Supplier at the address specified on the invoice. Unless otherwise agreed by the parties in writing, payment shall be made by Distributor by wire transfer in advance of shipment from the Supplier facilities. The Supplier Products shall be delivered EXW Supplier's facilities (Incoterms 2010). Risk of loss for the Supplier Products shall pass upon delivery to the named carrier at Supplier's facilities. If Supplier pays any shipping, insurance or handling costs, such costs will be billed to Distributor and will be reimbursed to Supplier by Distributor.

10. SECURITY INTEREST

Notwithstanding the passage of title, Supplier shall retain a security interest in all Supplier Products delivered until amounts for which Distributor is responsible under this Agreement have been received by Supplier. Supplier shall have all rights of a secured party, including the right to file a financing statement under the Uniform Commercial Code or a comparable law within the Territory to protect Supplier's security interest. In the event Distributor defaults in its payment obligations, Supplier will have the right to enter the premises of Distributor to recover possession of all Supplier Products on said premises, to recover all Supplier Products supplied by Distributor to its customers and associated supplies or software, and to pursue any other remedy existing at law or equity. Distributor, for itself and on behalf of its customers, hereby waives a prior hearing and demand for Supplier's exercise of such rights.

11. RESALE PRICE AND EXPENSES

Distributor shall set the selling price and license fees at which the Supplier Products are sold or licensed by it in the Territory. Distributor shall be solely responsible for the costs involved in the distribution of the Supplier Products, including sales costs, import duties, any and all banking charges, shipping and handling costs, installation costs or other operating expenses, letter of credit costs, wire transfer fees and other costs associated with making payment, and taxes, however designated, except that Distributor shall not be liable for taxes imposed that are based on Supplier's income.

12. PROMOTIONAL LITERATURE

Supplier agrees to furnish, in English, to Distributor (via email in pdf format) such descriptive literature, advertising materials, technical manuals and sales promotional materials concerning the Supplier Products as Supplier may, from time to time, have available for such purposes. Distributor shall have the right to translate such materials into the languages of the Territory at its own expense. Supplier shall retain ownership of all proprietary rights, including, intellectual property rights to the translated versions of the materials. Distributor will be solely responsible for the accuracy of the translations and will provide Supplier with a copy of each translated work. Distributor shall promptly revise (at Distributor's costs) the materials upon notice from Supplier.

13. USE OF TRADEMARKS

Distributor shall not be permitted to print, post or otherwise use letterhead, calling cards, literature, signage or other representations in the name of Supplier (or any of its affiliates) or to represent itself as Supplier (or any of its affiliates) or make commitments on behalf of Supplier (or any of its affiliates) without the express, written permission of Supplier. Distributor expressly agrees that no license to use Supplier (or any of its affiliates' trademarks, trade names, service marks or logos (collectively, the "Supplier Trademarks") is granted by this Agreement. Distributor may, however, indicate in its advertising and marketing materials that it is a distributor for Supplier Products and may, as necessary, incidentally use the Supplier Trademarks in its sales/marketing efforts. Upon request by Supplier, Distributor will place proper trademark, copyright and patent notices in its advertisements, promotional brochures and other marketing materials for Supplier Products. Supplier reserves the right to review Distributor's marketing and sales materials prior to their publication or use. No rights shall inure to Distributor as a result of any such use or reference, and all such rights, including goodwill shall inure to the benefit of and be vested in Supplier.

Upon termination of this Agreement for any reason, Distributor will immediately cease using the Supplier Trademarks as allowed in this Section and shall immediately take all appropriate and necessary steps to (a) remove and cancel any listings in public records, telephone books, other directories, remove any visual displays or literature at Distributor's location, the Internet and elsewhere that would indicate or would lead the public to believe that Distributor is the representative of Supplier (or any of its affiliates) or Supplier's (or any of its affiliates') products or services; and (b) cancel, abandon or transfer (as requested by Supplier) any product licenses, trade name filings, trademark applications or registrations or other filings with the governments of the Territory (whether or not such filings were authorized by Supplier) that may incorporate the Supplier Trademarks or any marks or names confusingly similar to the Supplier Trademarks. Upon Distributor's failure to comply with this paragraph, Supplier may make application for such removals, cancellations, abandonments or transfers in Distributor's name. Distributor shall render assistance to and reimburse Supplier for expenses incurred in enforcing this paragraph.

14. INFRINGEMENT BY THIRD PARTIES

Distributor will cooperate fully with and assist Supplier in its efforts to protect Supplier's intellectual property rights within the Territory and shall exercise reasonable diligence to detect and shall immediately advise Supplier if Distributor has knowledge of any infringement of any patents, trademarks, copyrights or other intellectual property rights owned or used by Supplier.

15. CONFIDENTIAL INFORMATION; NO REVERSE ENGINEERING

Supplier may provide Distributor with certain confidential or proprietary information ("Confidential Information"). Confidential Information includes information, whether written, electronic or oral, which Distributor knows or reasonably should know is proprietary, confidential or a trade secret of Supplier, including any and all technical or business information, the Software including its source codes and documentation, specifications and design information for the Supplier Products, servicing information, customer lists, pricing information, marketing information, policies, procedures and manuals regarding Supplier's distributors or distribution channels, research and development and other proprietary matter relating to the Supplier Products or business of Supplier. Distributor will refrain from using the Confidential Information except to the extent necessary to exercise its rights or perform its obligations under this Agreement. Distributor will likewise restrict its disclosure of the Confidential Information to those who have a need to know such Confidential Information in order for Distributor to perform its obligations and enjoy its rights under this Agreement. Such persons will be informed of and will agree to the provisions of this Section and Distributor will remain responsible for any unauthorized use or disclosure of the Confidential Information by any of them. Upon termination of this Agreement (or earlier, upon request by Supplier), Distributor shall cease to use all Confidential Information and promptly return to Supplier (or destroy, upon request by Supplier) any documents (whether written or electronic) in its possession or under its control that constitutes Confidential Information. During the term of this Agreement and thereafter, neither Distributor, nor Distributor's employees, independent contractors nor other agents shall (a) reverse engineer, decompile or otherwise disassemble the Supplier Products from the products themselves or from any other information made available to them, or (b) otherwise use any of the Confidential Information or Supplier provided training to support, maintain or otherwise service a third party's products or services.

16. COMPLIANCE WITH LAWS

In connection with its obligations under this Agreement, Distributor agrees to comply with all federal, state, local and foreign laws, constitutions, codes, statutes and ordinances of any governmental authority that may be applicable to Distributor, its activities under this Agreement or the Supplier Products, including all applicable export control laws and regulations. Distributor agrees to take all such further acts and execute all such further documents as Supplier reasonably may request in connection with such compliance.

17. PRODUCT WARRANTIES

(1) Limited Manufacturing Warranty. Supplier warrants for a period of [NUMBER] days following delivery of the Products (the "Warranty Period") that the Products shall be free from defects in materials and workmanship. Supplier's sole obligation under

this warranty shall be to provide, at no charge to Distributor, replacement Products. Defective Products must be returned to Supplier (at Distributor's cost) in order to receive warranty replacement (unless Supplier determines such return is not necessary) and shall become Supplier's property. For a warranty claim to be made, Distributor must follow the procedures established by Supplier from time to time.

(2) Warranty of Good Title. Supplier agrees to indemnify Distributor from any liability to any third party for infringement of United States patents, copyrights, trademarks or trade secrets with respect to Supplier Products sold/licensed by Distributor pursuant to this Agreement. This obligation does not extend to any foreign patents, copyrights, trademarks, or trade secrets or to any Supplier Products manufactured or modified by Supplier to meet Distributor's or a customer's specifications. Supplier shall, at its option, be allowed sole and exclusive control over the defense, settlement and compromise of any claims of infringement. Supplier must be notified in writing by Distributor within 10 days of any third party claim which, if upheld, might result in a liability subject to indemnification under this Subsection. If the distribution of the Supplier Products is threatened by a claim of infringement, or is likely to be enjoined or liability for infringement is found, Supplier may, in its discretion and at its sole option: (i) procure for Distributor the right to continue distributing the Supplier Products; or (ii) modify the Supplier Products so as to make them non-infringing; or (iii) substitute non-infringing products; or (iv) refund the price paid by Distributor for the Supplier Products in its possession subject to their return by Distributor and terminate this Agreement with respect to the allegedly infringing products. THIS SUBSECTION STATES THE ENTIRE LIABILITY OF SUPPLIER WITH RESPECT TO INFRINGEMENT OF ANY PATENT, COPYRIGHT, TRADEMARK, TRADE SECRET OR OTHER INTELLECTUAL PROPERTY RIGHT BY ANY SUPPLIER PRODUCT.

(3) Disclaimer. EXCEPT AS PROVIDED IN THIS SECTION, SUPPLIER MAKES NO OTHER WARRANTY, PROMISE OR OBLIGATION WITH RESPECT T O THE SUPPLIER PRODUCTS, THEIR USE, REPAIR OR PERFORMANCE. SUPPLIER DISCLAIMS ANY WARRANTY, PROMISE OR OBLIGATION THAT THE SUPPLIER PRODUCTS SHALL BE FIT FOR ANY PARTICULAR USE OR PURPOSE, REGARDLESS OF WHETHER SUCH USE OR PURPOSE IS MADE KNOWN TO SUPPLIER OR NOT. SUPPLIER DISCLAIMS ANY WARRANTY,

PROMISE OR OBLIGATION THAT THE SUPPLIER PRODUCTS CONFORM TO ANY SAMPLES OR MODELS. SUPPLIER HEREBY DISCLAIMS ALL OTHER WARRANTIES, PROMISES AND OBLIGATIONS, EXPRESS, IMPLIED OR STATUTORY, INCLUDING ANY WARRANTIES, PROMISES AND OBLIGATIONS ARISING FROM A COURSE OF DEALING OR USAGE OF TRADE. THE WARRANTIES SET FORTH IN THIS SECTION ARE INTENDED SOLELY FOR THE BENEFIT OF DISTRIBUTOR. ALL CLAIMS UNDER THIS AGREEMENT SHALL BE MADE BY DISTRIBUTOR AND MAY NOT BE MADE BY DISTRIBUTOR'S CUSTOMERS.

(4) Distributor's Warranties. Distributor agrees, at its cost, to provide its customers with, at a minimum, substantially the same warranties as set forth in Subsection 17(a). Distributor will assume all costs involved in providing any additional warranties.

18. REPORTING

Every month Distributor shall e-mail to Supplier a rolling 3 month, nonbinding sales forecast of the expected sales of Supplier Products in the Territory. In addition, Distributor shall furnish such other information in a timely manner in response to Supplier requests for information pertaining to Distributor's activities in the Territory. Such requests may include, prospect lists and status of prospect sales activity, information applicable to specific sales activities, data regarding competition in the Territory, product operational data and other information required by Supplier to effectively coordinate its international sales and marketing efforts.

19. INSPECTION OF RECORDS

Distributor shall keep accurate records of all its activities as reasonably necessary to determine its compliance with the terms and conditions of this Agreement, including accounting records, customer sales records and governmental filings. Distributor shall retain such records for at least a 3-year period following their creation or preparation. During the term of this Agreement and for a period of 18 months thereafter, Supplier shall have the right to inspect and audit such records.

20. TERM AND TERMINATION

Unless earlier terminated as provided in this Agreement, the term of this Agreement shall commence as of the Effective Date and shall automatically expire at the end of

[NUMBER] years following the Effective Date. Either party may terminate this Agreement as follows: (a) Immediately upon [NUMBER] days' prior notice with or without cause; (b) Immediately, for any breach or default of this Agreement by the other party which has not been cured within [NUMBER] days after the delivery of notice thereof to the party alleged to be in breach, specifying with particularity the condition, act, omission or course of conduct asserted to constitute such breach or default; (c) Immediately, upon the dissolution, insolvency or any adjudication in bankruptcy of, or any assignment for the benefit of creditors by, the other party or if the other party ceases to conduct business in the ordinary or normal course; (d) Immediately, if required by law or by any rule, regulation, order, decree, judgment or other governmental act of any governmental authority; or (e) Immediately by Supplier if Supplier reasonably suspects that Distributor breached any of its obligations of confidentiality or protection of Supplier's proprietary rights.

21. EFFECT OF TERMINATION

Upon notice of termination of this Agreement for any reason, the following provisions shall apply: (a) Supplier shall have the right to immediately appoint another distributor to serve existing customers and continue sales efforts in the Territory; (b) Supplier may continue to fill any orders from Distributor that have been accepted by Supplier prior to the termination of this Agreement under the terms and conditions of this Agreement; (c) All outstanding balances owed by Distributor to Supplier shall become immediately due and payable to Supplier; (d) Both parties shall at all times thereafter refrain from any conduct that would be inconsistent with or likely to cause confusion with respect to the nature of their business relationship; (e) All rights granted to Distributor under this Agreement shall cease, and where appropriate, revert to Supplier; and (f) Supplier, in its sole discretion, shall have the right, but shall in no way be obligated (unless otherwise required by law), to inspect and repurchase all or any quantity of the Supplier Products (including Supplier Products for demonstration and parts to service the Supplier Products) then owned or ordered by Distributor at the lesser of (i) the original price paid by Distributor for such Supplier Products, or (ii) at the then-current price to Distributor, and under both (i) or (ii), less any applicable restocking or refurbishing charge. Supplier shall have the right to assign such option to repurchase to any other person whom it may designate. No consideration or indemnity shall be payable to Distributor either for loss of profit, goodwill, customers or other like or unlike items, nor for advertising costs, costs of samples or supplies, termination of employees, employees' salaries and other like or unlike items. In no event shall Distributor continue to represent itself as a Supplier distributor or representative after termination of this Agreement.

Supplier shall have no liability to Distributor by reason of any termination by Supplier. Distributor shall indemnify and hold harmless Supplier from and against any and all liability, loss, damages and costs (including reasonable attorneys' fees) arising out of any claim by Distributor or any third party standing in the right of Distributor to any right of entitlement contrary to the express terms of this Section.

22. INDEMNIFICATION

Distributor agrees to indemnify and hold Supplier harmless from any and all acti ons, awards, claims, losses, damages, costs and expenses (including reasonable attorneys' fees) attributable to Distributor's breach of this Agreement or to any negligent, grossly negligent, willful or unlawful acts or omissions of Distributor, its employees, officers, agents, subcontractors, dealers or representatives.

23. RELATIONSHIP OF THE PARTIES

Distributor is an independent contractor and not an employee, agent, affiliate, partner or joint venture with or of Supplier. Neither Distributor nor Supplier shall have any right to enter into any contracts or commitments in the name of, or on behalf of the other or to bind the other in any respect whatsoever, except insofar as is allowed by this Agreement.

24. FORCE MAJEURE

Neither party shall be liable in the event that its performance of this Agreement is prevented, or rendered so difficult or expensive as to be commercially impracticable, by reason of an Act of God, labor dispute, unavailability of transportation, goods or services, governmental restrictions or actions, war (declared or undeclared) or other hostilities, or by any other event, condition or cause which is not foreseeable on the Effective Date and is beyond the reasonable control of the party. It is expressly agreed that any failure of the United States Government to issue a required license for the export of any Supplier Product ordered by Distributor shall constitute an event of force majeure. In the event of non-performance or delay in performance attributable to any such causes, the period allowed for performance of the applicable obligation under this Agreement will be extended for a period equal to the period of the delay. However, the party so delayed shall use its best efforts, without obligation to expend substantial amounts not otherwise required under this

Agreement, to remove or overcome the cause of delay. In the event that the performance of a party is delayed for more than 6 months, the other party shall have the right, which shall be exercisable for so long as the cause of such delay shall continue to exist, to terminate this Agreement without liability for such termination.

25. LIMITATION OF LIABILITY

SUPPLIER SHALL IN NO EVENT BE LIABLE FOR ANY INDIRECT, SPECIAL, EXEMPLARY, INCIDENTAL OR CONSEQUENTIAL LOSS OR DAMAGE OR FOR ANY LOST PROFITS, LOST SAVINGS OR LOSS OF REVENUES SUFFERED BY DISTRIBUTOR ARISING FROM OR IN ANY WAY CONNECTED WITH THIS AGREEMENT OR THE SALE, DISTRIBUTION OR USE OF SUPPLIER PRODUCTS. DISTRIBUTOR SHALL INDEMNIFY SUPPLIER AND HOLD IT HARMLESS FROM ANY CLAIMS, DEMANDS, LIABILITIES, SUIT OR EXPENSES OF ANY KIND ARISING OUT OF THE SALE, SUBLICENSE OR USE OF SUPPLIER PRODUCTS IN THE TERRITORY OR BY DISTRIBUTOR'S CUSTOMERS. THIS SECTION SHALL SURVIVE THE TERMINATION OF THIS AGREEMENT FOR ANY REASON.

26. GOVERNING LAW

This Agreement shall be governed in all respect by the laws of the State of [STATE], USA, which shall be applied without reference to any conflict-of-laws rule under which different law might otherwise be applicable. The United Nations Convention on Contracts for the International Sale of Goods shall not apply to any purchases or transactions entered into pursuant to this Agreement. Venue for any lawsuits brought by the parties to this Agreement against each other regarding or as a result of this Agreement shall be proper only in an appropriate [STATE] State Court or the United States District Court for the District of [STATE]. Distributor hereby submits itself to the exclusive jurisdiction of said courts and consents to service of process by confirmed facsimile transmission or commercial courier (with written verification of receipt returned to the sender).

27. ASSIGNMENT AND DELEGATION

Distributor shall have no right to assign any of its rights or delegate its obligations under this Agreement without the prior written consent of Supplier. Any assignment or delegation attempted without such written consent shall be void and of no legal effect whatsoever. This Agreement shall be binding upon the parties' respective successors and permitted assigns.

28. SEVERABILITY

In the event that any provision of this Agreement shall be unenforceable or invalid under any applicable law or be so held by applicable court or arbitration decision, such unenforceability or invalidity shall not render this Agreement unenforceable or invalid as a whole, and, in such event, such provisions shall be changed and interpreted so as to best accomplish the objectives of such unenforceable or invalid provision within the limits of applicable law or applicable court or arbitration decision.

29. CONSTRUCTION

The headings or titles preceding the text of the Sections and Subsections are inserted solely for convenience of reference, and shall not constitute a part of this Agreement, nor shall they affect the meaning, construction or effect of this Agreement. Both parties have participated in the negotiation and drafting of this Agreement. This Agreement is executed in the English language and may be translated into another language for informational purposes only. In the event an ambiguity or question of intent or interpretation arises, the English version of this Agreement shall prevail and this Agreement shall be construed as if drafted by both of the parties and no presumption or burden of proof shall arise favoring or disfavoring either party by virtue of the authorship of any of the provisions of this Agreement.

30. NOTICE

Any notice, consent or other communication required or permitted under this Agreement shall be written in English and shall be deemed given when (a) delivered personally; (b) sent by confirmed facsimile transmission; or (c) sent by commercial courier with written verification of receipt returned to the sender. Notice, consent or other communications (but not service of process) may also be given by e-mail. Rejection or other refusal to accept or the inability to deliver because of changed address or facsimile number of which no notice was given shall be deemed to constitute receipt of the notice, consent or communication sent. Names, addresses and facsimile numbers for notices (unless and until written notice of other names, addresses and facsimile numbers are provided by either or both parties) are provided below.

31. ENTIRE AGREEMENT; MODIFICATIONS; NO WAIVER; COUNTERPARTS AND SURVIVAL

This Agreement and the Exhibit attached hereto (which is specifically incorporated herein by this reference) contain the full and entire agreement between the parties with respect to the subject matter hereof. It supersedes all prior negotiations, representations and proposals, written or otherwise, relating to its subject matter. Any modifications, revisions or amendments to this Agreement must be set forth in a writing signed by authorized representatives of both parties. Distributor acknowledges and agrees that any failure on the part of Supplier to enforce at any time or for any period of time, any of the provisions of this Agreement shall not be deemed or construed to be a waiver of such provisions or of the right of Supplier thereafter to enforce each and every provision. This Agreement may be made in several counterparts, each of which shall be deemed an original. The provisions of this Agreement that, by express terms of this Agreement, will not be fully performed during the term of this Agreement, shall survive the termination of this Agreement to the extent applicable.

IN WITNESS WHEREOF the parties have caused this Exclusive Distribution Agreement to be executed and delivered by their duly authorized representatives.

[DISTRIBUTOR.Company]

__________Signature__________ ________Date________

[DISTRIBUTOR.First Name+Last Name]

[DISTRIBUTOR TITLE]

[SUPPLIER.Company]

__________Signature__________ ________Date________

[DISTRIBUTOR.First Name+Last Name]

[SUPPLIER TITLE]

常用句

1. We are interested in acting as exclusive agent for you in Taiwan.
2. We would like to offer you our services as a commission agent for Asia Pacific Area.
3. We shall act as sole agent for you and receive a commission of 5% on all sales in the whole territory of Taiwan.
4. If you would grant us an exclusive agent, we will make every effort to lead our business to complete success.
5. We shall be pleased to send you any further information you may need.
6. I hope to hear from you soon.
7. Please let me know if you need any further details.

練習題

1. 將下列的句子翻成英文

A. 本公司很有興趣成為貴公司在亞太區的唯一代理商。

B. 如果您能授予我們成為唯一的代理商，我們將竭盡所能地將事業作成功。

C. 我們想成為貴公司在台灣區的唯一代理商，收取所有銷售價格的 10% 佣金。

D. 感謝您 2 月 17 日的來函，授權我們在台灣代理您的商品。

E. 我們在貴國已有代理商。

F. 我們的商品已經全部被代理。

G. 萬一將來情況有變，我們會再與您聯絡。

H. 請告知你們是否有興趣和我們合作，並讓我們知道你們獨家代理的條件。

I. 為了使生意進展，我們同意承諾貴公司成為我方獨家代理，試驗期為一年。

2. 將下列中文書信翻譯成英文

我們從廣告上得知貴公司正在尋求代理，我們很有興趣想知道貴公司的條件。

我們具有多年的經驗，並且非常了解此地的市場，還有很好的銷售網絡，因此我們有信心可以將貴公司的商品在此地推銷得很好。

感謝貴公司的考慮，並等候儘快回覆。

3. 請就第 2 題回覆客戶，接受對方成為我們商品 PT-MEL 的代理。

4. 請寫一封婉拒對方請求代理的信函。

CHAPTER

16 其他商業信函

Other Business Correspondence

商業書信往來中，除了正常的商業流程中客戶與供應商的書信往來之外，也會發生其他的商業社交信函，例如：通知函 (notice)、祝賀函 (congratulations)、邀請函 (invitation)、慰問函 (sympathy) 或弔唁 (condolence) 等。本章就以上幾種較常發生的商業情形，舉例說明。

16.1 通知函

範例 16-1 通知公司地址變更

Dear Valued Clinet:

Please be advised that Taiwan Bear Corporation will be relocating on January 10, 2021. Please direct all correspondence to our new office at:

Taiwan Bear Corporation
358, Sec. 4, Xinyi Road,
Xinyi Dist., Taipei 11051
Taiwan

Our phone and fax numbers will remain the same:
Tel: 886-2-22584776
Fax:886-2-22589988

Thank you for your attention. We look forward to continued prosperity in our new location.

Sincerely,
Taiwan Bear Client Services

翻譯

我們在此要通知您，本公司台灣黑熊公司將於 2021 年 1 月 10 日搬遷。請於我們搬遷後將所有信件直接寄至我們的新辦公室：

台灣黑熊公司
11051 台北市信義區信義路 4 段 358 號，台灣。

我們的電話和傳真號碼維持不變：
電話：886-2-22584776
傳真：886-2-22589988

感謝知悉。我們期待在新的辦公室持續創造好的業績。

範例 16-2　通知公司合併

Dear Mr. Johnson:

As I am sure you are aware from recent news reports, First Bank and Dayton Credit Union are joining hands to better serve you. These two highly respected financial institutions officially merged on March 9. Both institutions will now operate under the name First Bank.

Perhaps you are wondering how this change affects you — our valued customer. There will be no immediate effects on your accounts regardless of the institution with which you were originally affiliated. In the near future, however, you will receive information concerning new banking policies which will include: personalized banking services, increased number of ATMs at convenient locations, and lower monthly service charges.

Please continue to address inquiries and business transactions to the usual account representatives. Our new team is looking forward to serving you more efficiently.

Sincerely,

翻譯

我相信您從最近的新聞報導中已得知 First Bank 和 Dayton Credit Union 將攜手合作，一起提供您更好的服務。這兩家備受尊敬的金融機構於 3 月 9 日正式合併。兩家機構合併後都將以 First Bank 的名義運營。

您或許會想知道我們合併後對客戶會有如何的影響。不管您最初所屬的機構如何，您的賬戶都不會立即受到影響。但是，在不久的將來，您將收到有關新銀行業務政策的信息，其中包括：個性化銀行服務，新增的自動提款機的放置地點及更低的每月服務費用。

請持續對本公司的支持。我們的新團隊期待能以更高的效率為您服務。

範例 16-3　通知結束營業

Dear Sirs:

This letter is to let you know that it has been a great pleasure doing business with you. We are closing the operation of our organization in this country and thus wish to bid you farewell.

You have been handling all the advertising and promotion work of our organization since we started our operations in this country. We have had great response due to your advertising work. But, in order to cut down the business cost and due to depression in market, we will have to close our operations in this country and shift to Japan where our headquarter is located. Thanking all your staff members for their co-operation and support.

Hope to work together with you in future.

Best regards,

翻譯

我們公司即將結束在貴國的營運，我們在貴國與您有生意往來的這一段時間，感到非常愉快。因此在結束營運之前，寫一封信函與您告知。

自我們在貴國開展業務以來，您一直協助我們公司處理所有廣告和業務推廣的工作。由於您的協助，我們得到了很好的回響。但是，為了因應市場蕭條和降低營業成本，我們將不得不結束我們在這個國家的業務，並將業務轉到我們在日本的總部。感謝你們所有員工的合作和支持。

希望將來能與您再一次合作。

▶ 16.2 祝賀函

範例 16-4　祝賀新公司成立

Dear John,

It has just come to our attention that you have lately opened your new European headquarters in Brussels. Congratulations on your bold venture. As you know, our company has had a long business association in the UK. We look forward to collaborating with you in your European venture.

Please let us know if we can be of any assistance to you. We will be delighted to help. We wish you the very best of luck and a prosperous future.

Best regards,

翻譯

我們最近發現您在布魯塞爾設置了新的歐洲總部。恭喜您再一次的成功。如您所知，我們的公司長期在英國有設置一個分公司。我們期待在歐洲與您的公司能一起合作。

請告知我們是否有可以為您提供任何協助的地方。我們會很樂意提供幫忙。預祝您業績長紅。

範例 16-5　祝賀升遷

Dear Mr. Larson,

Congratulations on your promotion to sales manager at Jupiter Systems!

We at Taiwan Bear have always appreciated your hard work and dedication to your job and to the relationship between our two companies. Your performance and commitment play a large part in our continuing business partnership.

We are very pleased at your success, and we are sure that Jupiter System has selected the best candidate for its Sales Department.

Best regards,

翻譯

祝賀您晉升為 Jupiter Systems 的銷售經理！

我們公司在台灣的同仁對於您一直以來的辛勤工作和對工作的奉獻，以及扮演兩家公司之間重要的橋梁表示非常感謝。您的績效和承諾在我們雙方的業務合作中扮演著重要角色。

對您的晉升我們替您感到非常高興，我們相信 Jupiter Systems 已經為其銷售部門選擇了最佳人選。

範例 16-6　祝賀獲獎

Dear Annie,

I came to know that you'll be awarded as the 'Employee of the Year' this year and I would like to express my congratulations to you. As a former team leader and co-worker, I know how much you deserve this award.

I remember the hard work and sincere efforts you always put in each and every task assigned to you. You have always finished and submitted your work on time and you always made sure to promptly submit your reports before the due date. I never experienced any problem when I was your team leader. Your good work ethics and attitude toward work which lifts up my spirit and motivates everyone to work hard as well. You are an asset to the company and it is recommendable to recognize you for your dedication and hard work.

Congratulations again and may you achieve more success in your career.

With warm regards,

我知道你今年得到「最佳年度員工獎」的榮耀，在此向你表達祝賀之意。作為你的前任主管和同事，我知道這獎項非你莫屬。

我記得你總是非常辛勤的工作並真誠的努力在每一項分派配給你的任務上。你總是按時完成並提交工作結果，並且你始終能確保在截止日期前準時提交你的報告。當我擔任你的小組組長時，我從未遇到任何業務上要解決的問題。你良好的職業道德和工作態度一直都能激發我的戰鬥力，並激勵每個人都努力工作。你是公司的重要資產，你獲得此獎是公司對你工作認真的感謝與認可。

再次恭喜你，並希望你的職業生涯上能取得更多的成就。

▶ 16.3 邀請函

範例 16-7　邀請客戶來看展覽

Dear Mr. Peterson,

It gives me great pleasure to invite you to the CES Show (Consumer Electronics Trade Show) in US, to be held in Las Vegas Convention Center from January 9, 2021 to January 12, 2021. For the convenience of everybody, the timings have been fixed from (11.00 a.m to 9.00 p.m)

The aim of the exhibition is to give a chance to everyone who is interested in latest technology to interact with one another.

We look forward to seeing you at the show.

Yours truly,

現在很多邀請函都是用 e-mail 的電子邀請函，而且會附註請對方回覆是否參加以利統計人數，在信件上會出現 RSVP 的字樣，是指請對方儘速回覆。

翻 譯

我們非常高興地邀請您參加 2021 年 1 月 9 日至 2021 年 1 月 12 日在美國拉斯維加斯會議中心舉行的美國消費電子展 (CES Show)。為了方便大家來看展，展出時間已經確定從上午 11 點到晚上 9 點。

此次展覽的目的是為所有對最新技術感興趣的人提供一次與其他廠商互動的機會。

我們期待著在展會上見到您。

範例 16-8　邀請客戶來看展覽

INVITATION

To whom it may concern:

Subject: China Import and Export Fair, Pazhou Complex
Booth No.: M200
Date: From Oct. 23 to Oct. 27, 2021

We are pleased to inform you that we will be exhibiting our newest products at the Canton Fair and would like to take this opportunity to introduce these products to you at a special price. Your visit to our booth will be our greatest honor. Please kindly confirm return:

☐ Will attend
☐ Will not attend

Best regards,

註 China Import and Export Fair Pazhou Complex 中國進出口商品交易會琶洲展館每年四月及十月的廣交會 (Canton Fair) 皆在此舉行。

Tips 邀請函可以用勾選的方式，請客戶於 email 的電子邀請函上回覆，一方面節省對方的時間，一方面有利於我方統計人數。

邀請函

敬啟者：
主題：中國進出口商品交易會，琶洲展館
展位號碼：M200
日期：2021 年 10 月 23 日至 10 月 27 日

我們很高興地通知您，我們將在廣交會上展示我們的最新產品，並希望藉此機會以特價推出這些產品。您到我們展出的攤位上來參觀將是我們最大的榮幸。請勾選下列並回覆以確認是否前來：

☐ 樂意前往
☐ 不克前往

範例 16-9　邀請客戶演講

Dear Mr. Peterson,

This is to formally invite you to lecture at our organization. The schedule we would like to propose is as shown below:

Date: March 15
Place: Johnson Auditorium
Attendants: About 100 newly hired employees
Objective: To promote a better understanding of fundamentals of FinTech.

I am looking forward to personally attending your lectures very much.

Yours truly,

翻譯

本信函是正式邀請您到我們的公司演講。以下是我們目前安排的時間表：

日期：3 月 15 日
地點：Johnson 大會堂
參與人員：大約 100 名新僱用員工
目標：加強對金融科技基礎知識的更佳認識

非常期待聆聽您的演講。

▶ 16.4 慰問函

範例 16-10　慰問病情

Dear Mr. Anderson,

I am surprised to hear that you are ill in bed and have not been able to work for several weeks. I am very concerned about your health and hope that you will recover very soon.

My colleagues join me in sending you our best regards and wish you a complete and speedy recovery.

Warm regards,

我很驚訝聽到了您生病的消息，而且已經幾個星期都無法工作了。我非常關心您的健康，希望您可以很快恢復健康。

在此我與我的同事們一起向您致上最誠摯的問候，並祝您很快就可恢復健康。

範例 16-11　慰問病情

Dear Mr. Anderson,

When I called at your office yesterday, I was very sorry to learn that you had been in a car accident on your way home from work recently. However, I was equally relieved to learn that you are making good progress and are likely to be back at work again in a few weeks.

Wishing you a speedy recovery!

Warm regards,

我昨天打電話到您的辦公室才得知您最近在下班回家的路上遇到了車禍。 還好我也同樣欣慰地得知，您目前恢復良好，並有可能在幾週內重新回到工作崗位。

祝您早日康復！

16.5 弔唁函

範例 16-12　弔唁客戶高階主管的噩耗

Dear Ben,

It is with great sorrow that we have received the sad news of the sudden and untimely passing of your CEO, Mr. Steve Jobs. We know how much he meant to you and your organization both as affine leader and as a friend.

Please accept our heartfelt condolence and convey our deepest sympathy to his family and associates.

Sincerely yours,

我們最近收到了貴公司執行長 Steve Jobs 先生突然過世的消息，感到非常悲傷。我們知道他對您和您的組織是多麼地重要，他扮演的不只是貴公司最重要的領導者也是如朋友一般的主管。

請接受我們衷心的哀悼，並向他的家人和同事表達我們最深切的慰問。

範例 16-13　弔唁客戶的親友噩耗

Dear Ben,

I have just learned with deep sadness of the death of your wife.

There is not much one can say at a time like this, but all of us at Taiwan Bear who have dealt with you would like to extend our sincere sympathy at your loss.

We all join in expressing our sympathy to you and your family at this very sad time.

Sincerely yours,

我剛剛得知您妻子過世的消息，對此深感悲痛。

在這樣的時候很多語言都無法表達我們對此的哀傷，但是我代表我們公司和所有同仁對您表達最深的慰問。

在這個令人悲傷的時刻，我們公司對您與您的家人致上最深的弔唁。

常用句

1. We would be delighted if you could attend the launch of our new product.
2. You are invited to attend the company's 10th annual formal party.
3. I would like to extend my warm congratulations on your promotion to manager of sales department.
4. We are delighted to learn that you have been awarded as Outstanding Employee of the year.
5. I have just learned with deep regret of the death of Mr. Jackson.
6. We were shocked to hear of the sudden passing of your CEO.
7. I will be calling on you in a little while to see if there is any way in which I can be of help.
8. Please convey our deepest sympathy to your family.
9. We all join in expressing our sympathy to you and your family at this very sad time.

練習題

1. 將下列的句子翻成英文

A. 對於您晉升為電子工業公司的董事，在此獻上誠摯的祝賀。

B. 我今早閱讀中國郵報時，看見您的名字列於新年榮譽榜上，我希望和大家一起向您道賀。

C. 得知 ××× 過世的消息，我感到相當難過。

D. 很高興宣布，我們 4 月 1 日已與 K&M 會計有限公司密切結盟。

E. 很高興向您宣布，我們的台北分公司將於 9 月 1 日開幕。

F. 史密斯夫妻誠摯邀請　參加小女瑪莉的婚禮。

2. 將下列中文書信翻譯成英文

邀請函

這是通知貴公司本公司將參加自 10 月 8 日至 10 月 15 日在台北世貿中心舉行的台北電子展，我們的攤位號碼是 A-016。

我們熱烈歡迎貴公司來參觀，展出後，我們也可以做進一步的生意討論。

當您確定拜訪行程時，請不要遲疑讓我們知道您的班機明細，以便安排人員到機場接機。盼望很快見到您。

3. 寫一封祝賀的信函，祝賀同儕在公司任職已十週年。表示公司的成功與成長都是仰賴像他這樣的員工，並且感謝員工的貢獻讓公司在這個產業上屹立不搖。希望大家繼續共事，並於這個特別的週年慶獻上恭賀之意。

4. 請寫一封通知函，告知客戶本公司於今年 7 月 1 日起將搬遷至台北市中心，請告知新地址與聯絡方式，並請客戶修正資料，於 7 月 1 日後將所有信件寄到新住址。

CHAPTER

17 求職信

Job Hunting

17.1 求職信

求職信是求職者寄給公司以應徵職務的信件。求職信通常被稱為「履歷封面頁」(cover letter)，因為在求職信中，應徵者需跟對方作簡略的介紹自己，並且推銷自已。一般求職信都會伴隨個人的簡歷 (resume/ curriculum vitae)。如果有以前主管或是學校師長的推薦信 (recommendation letter) 也應一併附上。

求職信寫的篇幅不宜過長，因為對方不會有太多時間瀏覽冗長的信件，因此要避免對方可能不感興趣的內容。信件中要避免概括的說法，要提出具體的資訊，例如：闡述自己重要而且出色的經歷與成就，像是得獎紀錄、難得的工作經驗或是其他成就等。

在應徵之前，應先對該公司和其職缺作事前的研究，找出該公司的經營理念和目標，然後讓對方知道你與他們有同樣的理念與目標。另外對職缺的資格要求與職務內容需詳細了解，並說明過往經驗曾負責過哪些類似的職務，藉此闡述自己的工作態度與風格。

範例 17-1　求職信：報章雜誌廣告得知消息

February 19, 2021

Jack Turner
Human Resource Manager
Pioneer Corporation
3080 Bowers Avenue
Santa Clara, CA 95054
USA

Dear Mr. Turner,

Subject: Application for the position of secretary

I was interested to see your advertisement in today's Daily Technology and would like to be considered for this position.

I have been Secretary to the Marketing Director at General Electronics for several years. As well so various administrative duties, I am required to attend and take minutes at meetings, deal with callers and correspondence in my boss's absence, supervise junior staff and other duties. I thoroughly enjoy my work and am very happy here. I do believe that my experience in secretary has prepared me for the responsibility of your open position.

Attached please find my resume for your further reference and evaluation. I believe that I can fulfill the requirements in your company, and hope I may be given the opportunity of an interview.

Sincerely yours,

Linda Chen

Linda Chen

Encl.

翻譯

主題：應徵秘書職位

我於今日在 Daily Technology 的雜誌上看到貴公司刊登的廣告，有秘書的職缺要找人，我想來應徵這個職務。

我在 General Electronic 公司擔任行銷處長的祕書已有多年。擔任此職務期間我參與了各種行政任務、紀錄出席的會議紀錄、在老闆不在的時候要處理來電和來信並且要指導初級員工和其他一些工作項目。我很喜愛我的工作，在這裡工作非常開心。我相信我在祕書方面的經驗已經為我準備好擔任貴公司這次職務要尋找的人。

附上我的履歷供您進一步參考和評估。我相信我能滿足貴公司的要求，並希望能有一次面試的機會。

範例 17-2　求職信：人力銀行得知消息

To	jackturner@pioneer.com
From	lindachen@gmail.com
Cc	
Bcc	
Subject	Application for Sales Assistant
Attachment	resume.doc

Dear Mr. Turner

I am glad to know from 104 job bank that you have a job vacancy for a Sales Assistant, and feel interested in applying for this position.

I have recently graduated from the Department of International Trade, Taiwan University and have had the training in International Trade Practice and Business English Communications. I am acquainted with MS Office and have some basic learning about ERP system developed by Oracle. Though I still have no any full-time working experience, I have the confidence to take the job well.

Enclosed please find my resume for your reference. I would highly appreciate if you could grant me an opportunity of interview as well as a possibility to work in your esteemed company.

Thank you very much for your consideration and look forward to hearing from you soon.

Best regards,
Linda Chen

很高興從 104 工作銀行得知，您有一個業務助理職務的空缺，並且對應徵這個職位有興趣。

我最近才從台灣大學國際貿易系畢業，並接受過國際貿易實務和商務英語方面的培訓。我熟悉微軟 Office 的作業系統，並對 Oracle 開發的企業資源整合系統 (ERP) 的基本知識有一定的了解。雖然我現在還沒有全職的工作經驗，但我有信心把這份工作做好。

隨函附上我的履歷供您參考。如果您能給我一個面試的機會，並且有機會在貴公司工作，我將非常感激。

非常感謝您的考慮，並期待儘快收到您的回覆。

範例 17-3　求職信：朋友處得知消息

To: jackturner@pioneer.com
From: lindachen@gmail.com
Cc:
Bcc:
Subject: Application for Sales Manager
Attachment: resume.doc

Dear Mr. Turner

Mr. William Naish, your Training Manager, has told me that your company has a vacancy for a Sales Manager. I would like to be considered for this position.

As you will see from my enclosed resume, I have 10 years' experience in the sales department of three well-known companies. My duties at Dajian Technology Ltd include training sales staff, dealing with the company's sales promotion programmers. I thoroughly enjoy my work and feel confidence that my experience in marketing has prepared me for the responsibility of full sales management.

Mr. Harrison Hu, Sales Director of my former company, has agreed to provide a reference for me. His details are shown on my resume. I shall be pleased to provide any further information you may need, and hope I may be given the opportunity of an interview.

Yours faithfully,
Linda Chen

翻譯

貴公司的訓練經理 William Naish 先生告訴我，貴公司有一個銷售經理的空缺。我想來應徵這個職位。

如您將從我附上的簡歷中看到的，我在三家知名公司的業務部門已有 10 年的經驗。我在 Dajian 科技有限公司的職責包括培訓銷售人員，處理公司業務的促銷工程，我非常喜歡我的工作，並且對我在市場行銷的經驗相當有信心可以使我管理好一個業務部門。

我前公司的業務處長 Harrison Hu 先生已同意為我推薦貴公司的此一職位。他的推薦說明附在我的簡歷上。我很樂意提供您可能需要的任何進一步信息，並希望我有一個面試的機會。

▶ 17.2 履歷表

一般美式履歷表稱為「resume」，英式履歷表稱為「CV」(curriculum vitae)。美式履歷一般只有一到兩頁，在結構上也比較彈性，通常著重於求職者的工作經驗與技術，還有教育背景。美式履歷幾乎都是按著欲應徵的職務量身訂做。美式履歷通常以倒敘方式列出資料（最近的事件先列出），例如：最後一個工作經驗先寫。

英式履歷會完整列出個人的經歷與成就，不管是教育上的、專業上的、志願的、榮譽的或是其他項目等等。因此英式履歷通常比美式履歷更長、更詳細。一般英式履歷會有兩頁左右，但是如果資料很多，也有可能長達好幾頁。傳統的英式履歷會按照時間順序，詳細列出相關資料，並且通常伴有封面頁，封面上會總結 CV 的重點，並強調求職者是該職缺的理想人選。

英式履歷應該要包含下列資訊：

1. 個人資訊（年齡／國籍／完整的連絡資訊）。
2. 教育背景（上過的學校與成績／學術獎項／研究經歷），從時間反序來寫，也就是最後取得的學位先寫出來，把重點放在最高的學位。
3. 工作經歷（列出你在社會上從事過的工作，無論有無支薪／可以包含義工或榮譽性質的工作經驗），目前常見的寫法是由時間反序來寫，也就是最後的工作經驗先寫出來。
4. 會員資格（學術協會／專業協會）。
5. 出版著作（論文／文章／書籍）。
6. 獎項（學術、專業或其他性質的獎項）。
7. 特別的資格認證或培訓。
8. 興趣（以與應徵工作相關之興趣為佳）。

範例 17-4　履歷表：樣式（美式）

Your Name

Address, City, ST ZIP Code |Telephone| Email

Objective

To get started right away, just click any placeholder text (such as this) and start typing to replace it with your own.

Education

DEGREE | DATE EARNED| SCHOOL

- Major: Click here to enter text
- Minor: Click here to enter text
- Related coursework: Click here to enter text

DEGREE |DATE EARNED| SCHOOL

- Major: Click here to enter text
- Minor: Click here to enter text
- Related coursework: Click here to enter text

Skills & Abilities

MANAGEMENT

- Think a document that look this good has to be difficult to format? Think again! To easily apply any text formatting you see in this document with just a click, on the Home tab of the ribbon, check out styles.

SALES

- Some of the sample text in this document indicates the name of the style applied, so that you can easily apply the same formatting again. For example, this is the List Bullet style.

COMMUNICATION

- You delivered that big presentation to rave reviews. Don't be shy about it now! This is the place to show how well you work and play with others.

LEADERSHIP

- Are you president of your fraternity, head of the condo board, or a team lead for your favorite charity? You're a natural leader – tell it like it is!

Experience

JOB TITLE |COMPANY| DATES FROM – TO

- This is the place for a brief summary of your key responsibilities and most stellar accomplishments.

JOB TITLE |COMPANY| DATES FROM – TO

- This is the place for a brief summary of your key responsibilities and most stellar accomplishments.

範例 17-5　履歷表：範例（美式）

Laurie Ann Wright
1234 West 5th Avenue
Seattle, WA 54321
(206)555-5555
lauriewright@gmail.com

OBJECTIVE

To obtain a position as a Legal Assistant with Legal Eagle Associates, to display excellent legal research and critical thinking skills in a criminal defense environment.

QUALIFICATIONS

Two years law office experience.
Analytical problem solver.
Excellent oral and written communication skills.
Ability to work well both cooperatively and independently.

WORK HISTORY

Legal Assistant ABC Lawyers Group October 2012-Present
Case organization including timeline management.
Initial case documents drafting and filing without supervision.
Jurisdiction and precedent legal research.
Wrote correspondence communicating with clients.

Legal Assistant Intern Freedom Law Firm September 2011- July 2012
Assisted with drafting and filing of legal documents with court clerk.
Created case files; managed paperwork and mailings.
Answered phone calls and client e-mails.
Scheduling of attorney meetings and client appointments.

EDUCATION
Associate of Applied Science in Paralegal Studies, 2011.
Washington University 123 Legal Lane, Seattle WA 54321
Graduated with 3.8 GPA.

REFERENCES

Gloria Alred Attorney at Law (555)555-5555	Robert Shapiro Senior Law Partner (555)555-5555	Mark Geragos Defense Attorney (555)555-5555

範例 17-6　履歷表：範例（美式）

PAMELA EVANS

145 Tahquitz Canyon - Palm Springs, California 92262 + 760-555-1212 - support@resumeedge.com

PROFILE

- Award-winning, multilingual Business Student with extensive professional and entrepreneurial experience.
- Awarded 2001 Student Leader for exemplary service in student government.
- Received 2001 Service Award for outstanding contributions to campus activities.
- Fluency in English, Spanish, and Portuguese. Technically proficient in MS Word, Excel, and Power Point; programming in Visual Basic and HTML; Wed design.

EXPERIENCE

TRANSLATOR, Orange County, California
Private Contractor……………………………………………………4/00-Present
- Team with two secretarial assistants to provide conversation-based translation and mediation services to non-English speaking business owners and employees.

Accomplishment:
- Awarded Hispanic Business Community recognition for assisting immigrants.

TTVXV, LLC, Santa Ana, California
Foreign Currency Trader, Intern….. ……………………………….12/01-2/02
- Handled $50,000+ monthly in trades and investment, specializing in Euros, Dollars, and Yen transactions; investigated trends and issued market reports.

Accomplishment:
- Increased profitability by exploiting Euro-to-Dollar exchange rate fluctuations.

CHILDREN'S LEARNING CENTER, Fullerton, California
Founder/ Business Manager……………………………………………...6/00-8/01
- Established and operated an educational institution with a staff of 20.

Accomplishment:
- Built revenues through direct student recruitment and cooperative local network.

PREMIER LEARNING ACADEMY, Irvine, California
Assistant Business Manager/ Spanish Tutor……………………………12/99-3/00
- Aided management and coached students for educational preparation institute.

Accomplishment:
- Boosted student enrollment; won Employee of the Month Award

SEDAY FIBRAS, S.R.L., Hemandarias, Alto Parana, Paraguay
Assistant Business Translator……………………………………………2/97-12/98
- Conducted English-Spanish – Portuguese translations of business documents and person –to – person conversations for global textile exporter.

EDUCATION & ACTIVITIES

CALIFORNIA COMMUNITY COLLEGE, Irvine, California
Business Administration Major, 2000-Present
- 4.00 GPA, President's List, Alpha Gamma Sigma, Phi Alpha Mu, Mu Alpha Theta.
- 2001 Associated Board of Trustees Member.
- Student Representative to Academic Senate, Spring 2001.
- Student Representative to Transfer Advisory Board, Spring 2001.
- Student Advisor to Business Club, Fall 2001.

範例 17-7　履歷表：樣式（英式）

Example of CV

Page 1/2

Name: CV writer
Address: Include full address and postcode
Tel: Include your cell phone and/or home phone
email: contact email address

PERSONAL SUMMARY

Use this section to sum up your education, work history and aims in a couple of sentences.
- e.g. "I am a hard working, reliable and motivated student with a strong science background. I am currently completing A-levels in Maths and Sciences and looking to gain experience working in engineering with the aim of training to become a cifil engineer".

MAIN ACHIEVEMENTS

- Make a bullet point list
- …of the main things you have achieved at school and in work. Include dates and company names.
- e.g. - Two weeks work experience with Madeup IT company – March 2017
- e.g. - Young enterprise Award – 2015
- e.g. – Captain of school football team – 2012-2013

SKILLS

List your personal, technical and specialist skills PLUS EXAMPLE. Don't just say "I have excellent communication skills", give people an idea of why.
- e.g. - Team work – Organized charity fashion show as part of school committee
- e.g. - Computing – Intermediate skills in MS Office, PHP, HTMLS
- e.g.- Communication – Work experience at reception in office, led school science research project.

WORK EXPERIENCE

Include: Job title, company, dates from – to and list what you did plus your responsibilities.
e.g. "Work experience, Madeup Company, March-April 2014 – Helped to update office contacts database, researched and wrote articles for website.
Remember to include voluntary work too!

Page 2/2

EDUCATION

List all your qualifications including GCSEs, A-levels, Higher, BTEC and short courses.

Put the highest level qualifications first plus results or 'TBC' if you are still studying.

e.g. A-levels: Maths - A, Physics - B, English - C.

GCSE: Math - B, Chemistry - A, Physics - C, English - B, French - C.

INTERESTS

Steer clear of talking about your social life and stick to hobbies like playing music, sport and personal projects. If you have won an award or certificate, mention that here or in Main Achievements.

REFERENCES

Make sure you have contact details for one/two people who can provide a reference. This should be someone who has managed you at work/work experience or, if this is your first job, a teacher or tutor. You don't need to include the details of your references on your CV, just write 'References available on request' here.

範例 17-8　履歷表：範例（英式）

Leah Norman

1 Main Street, New Cityland, CA 91010
Cell: (555)322-7337
example-email@example.com

Summary

Loss prevention Officer focused on safety and security management, as well as asset recovery. Skilled in problem solving and conflict resolution. Upholds high ethical standards and is honest with supervisors, employees, and customers at all times regarding suspicious events and security questions.

Highlights

- CCTV surveillance
- Interior and exterior patrol
- Background checks
- Shift work experience
- First Aid and CPR certified
- Report writing
- Crime prevention methods
- High level of integrity
- Observant
- Reliable

Experience

Grow Mart Garden Center　　February 2012 to Current
Loss Prevention Officer
New Cityland, CA

- Continuously patrol premises to secure patrons and property.
- Monitor hot spot areas where security coverage is limited to check for violators.
- Use security cameras to check on parking lot frequently.
- Periodically work in plain clothes to mingle with shoppers and prevent thefts.
- Directly pursue shoplifters and notify management and law enforcement of incidents.

Disneyland　　March 2005 to January 2012
Loss Prevention Officer
Anaheim, CA

- Maintained continuous visible presence to perform surveillance of guests and employees for theft.
- Detained suspects and investigated incidents promptly.
- Frequently testified in court as witness in trial cases.
- Gathered and prepared needed documentation to support charges and purse resolution.
- Answered park guest questions and directed to locations.
- Volunteered to take extra shifts during busy periods.
- Mentored new officers on protocols and surveillance techniques.

Davis Research　　July 2004 to February 2005
Administrative Assistant
Anaheim, CA

- Managed all business correspondence and document preparation.
- Answered in-person and telephone inquiries about services and projects.
- Drafted letters and reports for executive use.

Education

George Washington College　　2004
Associate of Applied Science: Administrative Studies
New Cityland, CA
Coursework in Business Writing and Administration
Savannah Marsh Academic Achievement Award

範例 17-9 履歷表：範例（英式）

Alison Connelly
15 Scouser Drive, Liverpool, Mersyside, LV3 5GT
Cell: 01601-322355
email:alisonc@gmail.com

PERSONAL PROFILE

I am a second year Business Management student at the University of Birmingham. I have developed excellent analytical and leadership skills through my degree, as well as key customer service and communication skills through my part time job at The French Shop. My determination and dedication is highlighted by my achievement of a black belt karate. I am now looking to further develop and use my skills in a year in industry placement, specifically in marketing.

EDUCATION

- BSC BUSINESS MANAGEMENT - FIRST YEAR AVERAGE: 68%
- September 2016 - Present
- 1st Year: Principles of Marketing (72%), HR(65%). analytical Techniques (68%).
- 2nd Year: International Marketing, Consumer Behaviour, Organisational Management.

NUNS MONK SCHOOL, WELWYN, HERTFORDSHIRE AL9 6NN
2012 - 2016

- A level Business Studies, Economics, General studies, Photography (AAAA)
- GCSE Maths, English, Science (AAB), 8 additional GCSEs grade C or above

WORK EXPERIENCE

THE FRENCH SHOP - Sales Assistant/Cashier
Sept 2016 - Present

- Serving customers and dealing with their requests.
- Cash handling on the checkout and in the cashing up of tills.
- Dealing with customer queries, customer complaints, and refunds on the Customer Service Desk.

PIES R'US. - Bakery Production/Sales Assistant
Sept 2000 - 2016

- Stock replenishment and packing pies.
- Conducting quality tests.
- Have to work efficiently and within a team to avoid a back log of stock.
- Serving customers and dealing with their enquiries, orders, and requests.

ADDITIONAL SKILLS AND INTERESTS
- Linguistics - can speak confidently in French and Spanish.
- Special Needs Assistant - I have had training during my secondary education, working with children aged between 10~13 years old with learning difficulties.
- Trained for 7 years in karate, qualified as a black belt.
- Sign Language - basic level of sign language.

REFREES
Kaite Longridge
The French Shop
Hastings
Relationship: Former Employer
Ph: (06) 844 8745

範例 17-10　履歷表：範例（英式）

PERSONAL DETAILS

Name: Lauren Smith
Address: 4a Westpoint Road, Havelock North
Telephone: (06) 834 2487
Mobile: 029-43212544
Drivers Licence: Current

AREER OBJECTIVE

Rehabilitation Coordinator

PERSONAL STATEMENT

I am an enthusiastic person committed to working as a health care profession. By continually enhancing my knowledge and skills I seek to provide the best quality service. I gain considerable satisfaction from empowering people by working alongside them to promote their holistic well-being.

QUALIFICATIONS

20xx Postgraduate Diploma - Rehabilitation Studies, Massey University

20xx Bachelor of Nursing - Eastern Institute of Technology, Taradale, NZ

20xx Registered Enrolled Nurse, Hawke's Bay district Health Board

PROFESSIONAL SKILLS

- Extensive experience in clinical assessment
- Excellent communication skills both oral & written
- Comprehensive experience in a wide range of clinical settings
- Proven ability to plan and allocate time efficiently
- Strong commitment to quality assurance
- Strong organizational ability as evidenced in former roles
- Commitment to health and safety requirements
- Comprehensive knowledge of disability issues
- Experience working cross culturally
- Good analytical and conceptual skills
- Ability to be creative when resolving complex issues
- Extensive experience working with multi-disciplinary teams and agencies
- Strong team player
- Flexible and adaptable
- Energetic
- Strong professional ethics

PERSONAL ATTRIBUTES

- Good sense of humour
- Hard working
- Enthusiastic

EMPLOYMENT HISTORY

5/20XX - 5/20XX **Rehabilitation and Assessment Centre, St Mary's Hospital, London**

Position: Nurse

Areas of Responsibility

- Maintaining health and wellbeing of clients
- Evaluating and implementing rehabilitation plans
- Liaising with multi-disciplinary team members
- Communicating with clients and relatives

10/20XX - 4/20XX **Devon Nursing Agency, London, England**

Position: Live in fulltime Carer for client and family

Responsibilities:

- Provide daily care and needs of a disabled client
- Support family members
- Liaise with multi-disciplinary team
- maintain cultural sensitivity
- maintain confidentiality

3/20XX - 9/20XX **Bay Home Support, Hastings, New Zealand**

Position: Client Service Officer

Responsibilities:

- Provide needs assessments and service coordination to individuals with disabilities
- Creating and facilitating respite care packages
- Assessing residential care subsidises for rest homes and the chronically ill
- Liaising with multi-disciplinary teams to plan and implement care packages
- Managing and prioritizing workload
- Contributing to the effective running of the Service
- Ensuring cultural sensitivity when working with clients and relatives
- Working within policy and standards to ensure quality assurance

Achievement:

- Developed the role of Liaison Office responsible for discharge planning in the Hospital Rehabilitation Unit.

3/20XX - 9/19XX **Hastings Memorial Hospital, Hastings, New Zealand**

Position: Enrolled/Registered Nurse

Duties:

- Maintaining effective clinical care including discharge planning
- Ensuring patient safety pre and post operatively
- Infection control
- Providing culturally sensitive care

INTERESTS

- Enjoy reading
- Listening to music
- Watching Netball
- Dancing
- Family life

REFEREES

Joy McMannus

Relationship: Former Manager

Rehabilitation and Assessment Unit

St Mary's Hospital

London

Ph: 00 44 878 5002

Colin White

Relationship: Former Manager

Bay Home Support

Hawke's Bay Hospital

Hastings

Ph: 06-878-88952

17.3 推薦信

應徵者在請人寫推薦信有兩種型態，一種是就職證明書 (employment reference)，另一種是推薦信 (letter of recommendation)。兩種都是寄給應徵者的未來雇主，但是寫的方式與目的稍有不同。

就職證明書是證明應徵者曾於該公司工作，並且就職證明書並不是寫給特定的某人，所以書信應該簡短客觀，有時甚至是應徵者自己書寫，再請推薦人簽名。一般就職證明書應含下列事項：

1. 信中應註明應徵者與推薦人的關係（例如：主管）。
2. 應徵者於公司的就職期間。
3. 應徵者於公司的職務與工作內容。
4. 應徵者的工作態度和個人特質。

推薦信則通常是由應徵者的前主管或是學校的老師、教授針對應徵者所求，針對特定對象或是公司推薦應徵者的信函，因此在信函中要闡述應徵者的良好工作習慣、技巧、才能和學術上的傑出表現等正面特質來做強力的推薦。

範例 17-11　就職證明書（給不特定人士）

To whom it may concern:

I confirm that Sophia Brooks was employed as a sound engineer with Electronic Music from July 2017 to May 2021.

As a sound engineer, Sophia was responsible for conferring with performers and producers to achieve the desire sound. She tested and maintained our mixing and recording equipment and recorded, mixed, edited, and reproduced a variety of vocals, music, and sound effects. She is skilled at regulating sound quality during recordings. She also showed herself to be dependable, reliable, and creative in problem solving.

Sophia was considered a valuable member of the team who consistently achieved good results and met all expectations.

Sincerely yours,

Linda Chen

Linda Chen
Manager

在此證明 Sophia Brooks 於 2017 年 7 月至 2021 年 5 月於 Electronic Music 受雇擔任音響工程師。

Sophia 擔任音響工程師時，負責與表演者和製作人進行協商，以得到預期的音效。她必須測試並維護我們的混音和錄音設備，並錄製，混合，編輯和重新製作各種歌唱、音樂及音響效果。她尤其擅長在錄音過程中調整音響品質。在解決問題方面，她展現出獨立、可靠與創意的特質。

Sophia 在我們團隊中一直是一位不可多得的成員，她持續令人滿意的表現，並達到我們所有的期望。

範例 17-12　就職證明書（給特定人士）

Dear Mr. Roberts,

I am writing to you regarding the job application of Ms. Sheila White in your company.

Ms. White is known to me for the past four years. She was working in the capacity of 'Insurance Sales Executive' in our company. After working with Sheila for four years, I can definitely write few good lines about her capabilities and skills.

Ms. White's job profile was to market and sell our insurance policies. She was selling both business and personal policies. She was a great employee who undertook and executed all tasks I gave her very well. She would explain each and everything about an insurance policy to clients in a very friendly manner. She was very good at sales and sold hundreds of policies during her tenure with us and we were really satisfied with her performance. And, she would also take care of post sales customer service on her own.

I would say that Ms. White is a very talented, team-player and, hardworking and a sincere person. Undoubtedly, she will be an asset to any company. Should you need any additional information about Ms. White, please feel free to contact me.

Best regards,

Richard Clifton

Richard Clifton
Marketing Manager

翻譯

我寫這一封信給您是為了推薦 Sheila White 女士去應徵貴公司的工作。

我認識 White 女士已有四年的時間。她在本公司擔任「保險銷售主管」一職。在與 Sheila 合作四年之後，我深知她的工作能力和技巧。

White 女士的工作範圍包括推廣和銷售我們的保險單。她的業務包含銷售商業和個人的保險。她是一位非常優秀的員工，對於我所派任的工作，她一向執行的非常成功。她會以非常友好的態度向客戶解釋每一項有關保險單的事項。她在銷售方面非常出色，在她任職期間賣出了數百張保單，我們對她的表現非常滿意。而且，她還會自己提供客戶售後服務。

在此我想說，White 女士是一位非常有才華，具團隊合作，勤奮和真誠工作的人。毫無疑問，她將成為任何公司的資產。如果您需要關於懷特女士的其他訊息，請隨時與我聯繫。

範例 17-13　推薦信（教授推薦信）

Dear Mr. Carton,

This is in reference to Mr. Melvin Hamilton's job application in your company for the post of Jr. Engineer.

Melvin was a student of our institute. He was a meritorious student of our college. He excelled in all his semesters and ranked 5th. Further, I would like to say that his financial condition is not stable and he needs a job. But, this is not the only factor for recommending him, he is a brilliant science student and thus, I support his application.

I know about your organization and its involvements. 'New Engineering Solutions' is a prestigious company. It encourages new and innovative minds. I am recommending Melvin's candidature for this open position in your company. I am very much sure that being an intelligent person, he will definitely fulfill your expectations and will give his best. He is a sincere and disciplined person.

So, please consider his application for this position. Thanking you for your consideration.

Sincerely yours,

Rick Burton

Dr. Rick Burton
Professor
Jefferson Science College

翻譯

本信函是推薦 Melvin Hamilton 先生應徵貴公司的助理工程師的職位。

Melvin 是我研究所的學生，他在我們學院是一位優秀的學生。他在每一個學期都表現出色，並名列前茅。此外，我想特別說明的是他的財務狀況並不是很好，因此他急需一份工作。但是，這不是推薦他理由之一。我推薦他是因為他是一位非常出色的理工學生。

我知道貴公司是一家享有盛譽的公司。貴公司鼓勵新的創意和創新的的想法。我推薦 Melvin 去爭取貴公司的這一職位。我非常確定，作為一名優秀的學生，他一定會實現你的期望，並且會盡全力。他是一個真誠和紀律嚴明的人。

所以，請考慮他對這個職位的申請。感謝您的考慮。

範例 17-14　推薦信（主管推薦信給特定人士）

Dear Mr. Fox,

This is in reference to the application of Mr. Ben Carter in your Newspaper for the post of 'Sub-Editor'.

Mr. Carter has worked under me as a 'Proof Reader' for two years. I am writing in support of his job application. I would like to tell you that Ben is a very sincere person. He is very intelligent and has thorough knowledge about editing. I was always satisfied with his work. He excelled my expectations. He was hired as an intern, and he got promoted to 'Proof Reader' within a short span of time.

So, I refer Ben's application for the open position of 'sub-editor' in your paper. It would be really pleasing to see him work as a sub-editor.

I am very much sure that he would add value to your company.

Yours faithfully,

Jim Carpenter

Jim Carpenter
Asst. Editor
New Sun Times

本信函是推薦 Ben Carter 先生應徵貴公司的副編輯的職位。

Carter 先生已經在我部門工作已有二年，他所擔任的職務為「確認讀者」。我毫無保留推薦他申請貴公司的工作。我想說的是 Ben 是一個非常真誠的人，他非常聰明，對編輯有深入的了解，我對他的工作總是很滿意，他的表現超出了我的期望。他之前被本公司聘為實習生，並在短時間內就晉升到「確認讀者」的職位。

所以，我極力推薦 Ben 去申請這一職位。如果能看到他爭取到這個副編輯的職位，這將是令人感到高興。

我非常確定他會為貴公司帶來極大的利益。

範例 17-15　推薦信（主管推薦信給不特定人士）

To whom it may concern:

Miss Michelle Tan was employed as department assistant in this company's Sales Department when she left secretarial college in May 2016. She was promoted to my Personal Secretary in 2018 until she left the company in March 2021.

Her responsibilities included the usual secretarial duties involved in such a post as well as attending meetings, transcribing minutes and supervising and advising junior secretaries.

Michelle used her best endeavors at all times to perform her work conscientiously and expeditiously. She was an excellent secretary, and extremely quick and accurate shorthand typist, and meticulous in the layout, presentation and accuracy of her work. I cannot overstress her exceptional work rate which did not in any way detract from the very high standards she set for herself.

Michelle had a great working attitude, and she made a point of building great team relationships. It was a great loss to both myself and the company when Michelle took up teacher training.

In my opinion, Michelle has the necessary character, dedication and approach to be suitable for the position of personal secretary or to enter the teaching profession. I can recommend her highly and may be contacted for further information.

Sincerely yours,

Linda Chen

Linda Chen
General Manager

翻譯

Michelle Tan 小姐於 2016 年 5 月從秘書學院畢業後就在我們公司的業務部門擔任部門助理一職，並且於 2018 年晉升為我的私人秘書直至她於 2021 年 3 月離開本公司。

她的職責包括一般秘書應擔任的業務範疇，像是出席會議，抄錄會議記錄，監督和教育初級秘書等。

Michelle 總是竭盡全力並盡自己最大的努力去迅速地完成她的工作。她是一位出色的秘書，並且可以非常快速準確的速記，而且在工作上展現出嚴謹與準確。我無法再進一步強調她的工作效率，因為她對自己設定的標準就很高。

Michelle 有非常好的工作態度，她建立了良好的團隊關係。在 Michelle 去受教師培訓時，對我和公司都是一個巨大的損失。

我認為，Michelle 具備了擔任個人秘書或是教師行業的必要特質，那就是奉獻的精神。我在此極度推薦她，如果您需進一步的訊息，您可與我聯繫。

常用句

求職函：

1. I was interested to see your advertisement in today's China Post and wish to apply for this post.
2. I am writing to inquire whether you have a suitable vacancy for me in your organization.
3. I have learned from the internet that there is a vacancy in your sales department.
4. I am pleased to offer myself as the most suitable candidate for the position as a secretary of director.
5. I understand from Mr. Jones, one of your suppliers, that there is an opening in your company for graphic designer.
6. Please consider my qualification for the position of Electronic Engineer.
7. I hope you will consider my application favorably.

8. I look forward to hearing from you and to being given the opportunity of an interview.
9. Your kind consideration of granting me an interview would be grateful.
10. I look forward to the opportunity of attending an interview when I can provide further details.

推薦函：

1. I am only too happy to recommend Ms. Jane White for the position of sales.
2. I can wholeheartedly recommend Ms. Jane White for this position.
3. In my years of working with Mr. Ben Hope, he has shown himself to be a capable manager in Sales Department.
4. Mr. Ben Hope has been employed by this company from January 1, 20xx to December 31, 20xx.
5. I have pleasure in recommending Mr. Ben Hope highly and without hesitation.
6. I hope that Mr. Ben Hope meets with the success we feel he deserves.
7. I shall be sorry to lose Ben, but realize that his abilities demand wider scope than are possible at this company.
8. I can recommend Ms. Jane White to you with every confidence.

練習題

1. 請依報上的求才廣告，應徵銷售人員一職，撰寫一封求職信函。
2. 依自己的背景、特色，撰寫一篇履歷表。

▶ Appendix 附錄

Ⅰ. 商用英文常用字與專有名詞

常用字

A	
Acceptance	承兌
Advance freight	預約付費
Agency agreement	代理合約
Act of god	天災
Act of war	戰爭
Air freight forwarder	空運貨物承攬運送人
Air waybill (AWB)	空運提單
Arrival notice	到貨通知書
B	
Back date B/L	提單上簽發的裝船日期比實際裝船日提前一日以上的提單
Bill of sufferance	貨物保稅單
booking note	托運單
bonded warehouse	保稅倉庫
bulk cargo	散裝貨
C	
cargo	由船舶或飛機運輸之貨物
carriage	運送
carrier	運送人才船東或其船務代理公司
carrying vessel	載運船
Cash on delivery (COD)	到貨付現
combined transport document (CTD)	聯合運送單據
country of origin	原產國
cubic meter (CBM)	立方公尺 (M^3)
cubic foot/feet (cu.ft.)	立方呎（才）
currency surcharge	貨幣貶值附加價
D	
drawback	退稅
duty-paid	完稅價格

E	
Estimated time of arrival (ETA)	預計抵達時間
Estimated time of departure (ETD)	預計離開時間
Estimated time of sailing (ETS)	預計離航時間
F	
FedEx	美國聯邦快遞
Free trade zone	自由貿易區
I	
International Chamber of Commerce	國際商會
L	
Last shipping date	最後裝船日
Letter of indemnity (L/I)	拖運人繳交給船公司的補償書
Letter of guarantee (L/G)	保證書
N	
Negotiation	押匯
Not negotiable	禁止背書轉讓
O	
On board date	裝船日
P	
Port of loading	裝貨港
Port of discharging	卸貨港
Port of destination	目的港
S	
Sea-air combined transport	海空聯運
Seal	貨櫃封條
Shipper	拖運人
Short shipment	裝載短少
Shut out	退關
Specification	規格
Stamp duty	印花稅
Statement of merchandise	商品內容說明書
Survey	檢定
Surveyor	公證人
Survey report	公證報告書
T	
Tare weight	空貨櫃的重量
Tariff	運費表，關稅
Terminal handling charge	貨櫃場吊櫃費

To order	依持有人指示交付貨物
To order of shipper	依托運人指示交付貨物
To order of the buyer	依買方指示交付貨物
Trading promotion fee	推廣貿易服務費
Transshipment	轉運
Triangular trade	三角貿易
U	
Uniform rules and practice for Documentary credits (UPC)	信用狀統一慣例
Uniform rules of collection (URC)	託收統一慣例

專有名詞

Agent 代理商	
shipping agent	船務代理人
claims agent	索賠代理人
commission agent	委任代理人
forwarding agent	轉運代理人
purchasing agent	採購代理商
Bill of Lading (B/L) 提單，載貨證卷	
clean B/L	清潔提單
ocean bill of lading	海運提單
negotiable bill of lading	可轉讓提單
ocean bill of lading	海運提單
on board B/L	裝運提單
received B/L	備運提單
stale B/L	陳舊提單
through B/L	聯運提單
Certificate 證書	
export inspection	輸出檢驗合格證明書
export license	出口許可證
import license	進口許可證
insurance certificate	保險證明書
certificate and list of measurement/ weight	容積重量證明書
certificate of origin	產地證明書
Payment Terms 付款條件	
L/C: Letter of Credit	信用狀
COD: Cash on Delivery	交貨付現

CWO: Cash with Order	下單付現
B/C: Bill for Collection	託收票據
D/P: Document against Payment	付款交單
D/A: Document against Acceptance	承兌交單
O/A: Open Account	記帳，月結
D/N: Debit Note	應收帳款
C/N: Credit Note	應付帳款
Payment Method 付款方法	
T/T: Wire Transfer	電匯
M/T: Mail Transfer	信匯
D/D: Demand Draft	即期匯票
Letter of Credit 信用狀	
Back to back L/C	轉開信用狀
Confirmed credit	保兌信用狀
Sight L/C	即期信用狀
Stand by credit	擔保信用狀
Straight credit	直接信用狀
Transferable L/C	可轉讓信用狀
Usance L/C	遠期信用狀
Collection 託收	
Document against payment (D/P)	付款交單
Document against acceptance (D/A)	承兌交單
container freight station	貨櫃集散站
container ship	貨櫃船
container terminal	貨櫃碼頭
container yard	貨櫃場
Draft 匯票	
bearer	持票人
drawee	被發票人，付款人
drawer	發票人
payee	受款人
payer	付款人
Copies 份數	
Original	正本
Duplicate	二份
Triplicate	三份
Quadruplicate	四份
Quintuplicate	五份

Sextuplicate	六份
Septuplicate	七份
Octuplicate	八份
Nonuplicate	九份
Decuplicate	十份
Freight 運費	
freight collect	運費到付
freight conference	運費同盟
freight forwarder	貨物承攬業者
freight prepaid	運費已付
freight tariff	運價表
Insurance 保險	
Insurance policy	保險單
insurer	保險人
the insured	被保險人
riot and civil commotion	暴動與內亂
average statement	海損理算書
cover note	承保通知書
marine insurance	海上保險
open policy	預約保險單
premium	保險費
Institute cargo clauses (ICC)	協會保險條款
WPA: with particular average	水漬險
F.P.A.: free of particular average	平安險
A.R.: All risks	全險
General average	共同海損
Particular average	單獨海損
Total loss	全損
W.R.: War Risk	兵險條款
Strikes clause	罷工條款
Theft pilferage and non-delivery (Clause (T. P. N. D.))	失竊條款
Packing unit 包裝單位	
Ampoule	筒
Ball	粒
Bale	捆
Ballot	梱
Bar	條

Barrel	桶
Basket	籠
Bundle	束
Cake	塊
Card	片
Coil	捲
Dozen	打
Gallon	加侖
Gross	籮
Lot	堆
Pack	包
Piece	件
Set	套
Packaging 包裝物	
Bagging	袋裝
Basket	籃
Bottle	瓶
Box	盒，箱
Can	罐頭
Carton	紙板盒
Carton box	硬紙盒
Case	箱
Cask	桶
Corrugated paper box	瓦楞紙盒
Crate	板條箱
Drum	圓桶
Fiber board	纖維板
Glass	玻璃瓶
Gunny sack	麻袋
Jar	罐、壺
Padded bag	特製包裝袋
Paper thimble	紙殼筒
Straw bag	草袋
Tin	錫罐
Wooden case	木箱
Packing materials 包裝材料	
Box strapping steel	打包鐵皮
Paper scrap	紙屑

Paper wool	紙條
Saw-dust	木屑
Tin lining	鉛皮箱胎
Mark 標識；頭	
Triangle	三角形
Square	四方形
Pentagon	五角形
Hexagon	六角形
Heptagon	七角形
Octagon	八角形
Nonagon	九角形
Regular decagon	十角形
Asterism	三星形
Annulus	環形
Circle	圓圈
Concentric circle	同心圓形
Cross	十字形
Cross circle	橫雙圈
Cross in circle	圈中十字
Diamond	鑽石形
Double triangle	雙三角形
Ellipse	橢圓
Heart	心形
Lozenge	菱形
Oval	橢圓形
Quadrangle	長方形
Rhomboid	斜方形
Sector	扇形
Semi-circle	半圓形
Solomon Seal	六角星形
Star	星形
Shipping mark 裝運標識	
Main mark	主標識
Side mark	側麥頭
Case mark	箱號標識
Care mark	注意標識
Destination mark	收貨地標識
Export mark	出口標識

Port of loading	裝運港標識
Port mark	目的港標識
Quantity mark	數量標識
Warning mark 小心標識	
Dangerous	危險品
Explosive	易爆品
Fragile	易碎品
Inflammable	易燃品
Handle with care	小心搬運
Keep cool	保持涼爽
Keep dry	保持乾燥
Keep flat	保持平放
Keep from heat	勿近熱
Keep upright	切勿上下倒置
No hooks	勿用鉤
Poisonous	有毒的
This side up	此端向上
Price terms 價格條件	
Ex-factory/Ex-work	工廠交貨價
FAS: Free Alongside Ship	船邊交貨價
FOB: Free on board	船上交貨價（離港價）
FOR: Free on Railroad	火車上交貨
FOT: Free on Truck	卡車上交貨
C&I: Cost and Insurance	含保險在內價
C&F: Cost and Freight	含運費在內價
CIF: Cost, insurance, and freight	含運費、保險在內價
CPT: Carriage Paid to	包括成本、運費和保險支付至所指交貨地點價
CIP: Carriage and Insurance Paid to	包括成本和運費支付至所指交貨地點價
DAF: Delivered At Frontier	邊境交貨價
DES: Delivered Ex Ship	貨物運至到岸船內價
DEQ: Delivered Ex Quay	貨物運至目的港碼頭價
DDU: Delivered Duty Unpaid	貨物運至目的地稅未付訖價
DDP: Delivered Duty Paid	貨物運至目的地稅訖價
Shipping 航運	
shipping advice	裝運通知
shipping agency	船務代理
shipping company	船運公司

shipping order (S/O)	艙位訂單
shipping schedule	船期表
shipping space	艙位
shipping instruction	裝船指示
CY: Container Yard	整櫃
FCL: Full Container Load	整櫃
CFS: Container Freight Station	併櫃
LCL: Less than container load	併櫃
ETD: Estimated time of departure	預計離港日
ETA: Estimated time of arrival	預計到港日
D/O: Delivery Order	交貨單（小題單）
S/O: Shipping Order	裝貨單
S.S.: Steam Ship	汽船
Weight & Measurement 度量衡	
Actual tare	實際皮重
Area	面積
Bulkiness	笨重
Capacity	容積
Cubic	立方
Gross weight	毛重
Length	長度
Measure	度量
Metric ton	公噸
Net weight	淨重
Square	平方
Tare	皮重
Volume	體積
Weight ton	重量噸

Ⅱ. 電郵中常見的商業縮寫語

縮寫	全文	中文
A		
A.R.	against all risks	全險
A.B.NO.	accepted bill number	承兌匯票號碼
abt	about	有關
A/C	account	帳戶
AC	acceptance	承兌
A.D.	adjustment reversal	歸原處置
a/d	after date	票後限期付款
adv	advice	通知
ad val.	ad calorem (according to value)	以價值計稅
AFAIK	as far as I know	據我所知
AFAIR	as far as I remember	記憶所及
AG	Account General	總帳
AGRMT	agreement	協議
Agt.	agent	代理
a.k.a	also known as	別名，又名
AM	amendment	修正（信用狀）
A.M., a.m.	ante meridiem (before noon)	上午
Amt.	amount	金額
A.M.T.,M/T	Air Mail Transfer	信匯
AN.	arrival notice	到貨通知
a/o	account of	（某人）的帳
A/P	account payable or Authority to Purchase	應付帳款或是委託購買
Appln	application	申請
approx	approximately	大約，概略
APRV	approve	批准
A.R.	all risks	全險
Art.	article	條款
A/S	account sales	銷貨清單
a/s	after sight	見後限期付款
ASAP	as soon as possible	儘快
asstd.	assorted	各色具備的
ATTN, att., attn.	attention	注意

av., a/v	average	平均
A/V	according to value	計算價值
B		
B/-, b/-	bale, bag	捆、包
B4	before	以前，在…之前
B4N， BFN， BBFN	by for now	再見
bal., balce	balance	餘額
bbl.	barrel	桶
BBS	Bulletin Board System	電子壁報板
B/C	bill for collection	託收票據
bcc	blind carbon copy	密本
B/D	bank draft	銀行匯票
b'dle, bdl.	bundle	束
B/E	bill of exchange	匯票
bg	bag	袋
biz	business	生意，業務
BK	because	因為，由於
B/L	Bill of Lading	提單
B/N	bank note	銀行紙幣
b.o.	buyer's option	買方有選擇權
bot.	bottle	瓶
B/P	bills payable or bill purchased	買入票據出口押匯 應付票據
B/R	bills receivable	應收票據
brl., barl.	barrel	桶、樽
B/S	balance sheet	資產負債表
BTW	by the way	再者，題外話
bsh, b/s	bushel, bales	英斗、捆
bx	box	盒、箱
BYKT	but you know that	可是你知
C		
c	cent, centigrade	分，度數
C.A.	credit advice	信用通知
c.a.d.	cash against documents	依憑單付現
c.a.f., C&F	cost and freight	運費在內價
canc	cancel	取消
cat.	catalog	型錄

C.B., C/B	Clean Bill	光票
CBD	cash before delivery	交貨前付款
c.c.	carbon copy, cubic centimeter	副本，立方公分
C/C, C.C.	Chamber of Commerce	商會
C/D	Cash against Document	憑單據付款
C&D	collection and delivery	收款交貨
cert.	certificate, certified	證明書，證明
C.F.S.	container freight station	貨櫃集散場
C.F.&I	Cost, Freight, and Insurance	運費、保險費在內價
C.H.	clearing house	清除倉庫
chk, chq	check, cheque	支票
C/I	certificate of insurance	保險證書
C&I	Cost and Insurance	成本及保險單
c.i.a.	cash in advance	預付現金
C.I.F.	Cost, Insurance and Freight	成本、保險及運費
C.I.F.C.	Cost, Insurance, Freight and Commission	運費、保險費、佣金在內價
C.I.F.E.	Cost, Insurance, Freight, and Exchange	成本、保險、運費及匯兌價
CK	check	支票
ck, csk.	cask	樽
CL	collection	收帳
CM	commission	佣金
C/N, C.N.	Credit Note, Consignment Note, covering note	貸方通知單，發貨通知單，保險承保單
C.O.	Certificate of Origin	產地證明書
c/o	care of, carried over	煩轉，轉自
Co., CO.	company, conference	公司
c.o.d.	Cash on Delivery	交貨付款
Con. Inv.	Consular Invoice	領事發票單
CONF	confidential	機密的
corp.	corporation	公司，法人
C/P	charterparty	雇船租約
cr., Cr(s).	credit(s), creditor(s)	信用、貸方，債權人
C/S, cs	case of cases, case	箱
CU	see you	再見
CUL	see you later	再見，再聯絡
C.W.O.	cash with order	現金訂貨

cwt.	hundredweight	百磅重量單位
C.Y., CY	container yard, calendar year	貨櫃集散場，日曆年度
D		
D/A	Documents against Acceptance	承兌交單
D.A.	debit advice	借方通知
d/a	days after acceptance	承兌後…日付款
dba	do business as	從事如下工作
D.D.	demand draft	即期匯票
d/d	days after date	出票後…日付款
dept.	department	部門
diff.	difference	差額，不同
Disc	discount	折扣
DL, D/L	download	下載
D/N	debit note	借方通知
D/O, do.	delivery order	交貨
doz.	dozen	打
D/P	Documents against Payment	付款通知
DR, Dr(s).	debit(s), debitor(s)	借方，債務人
d/s., d.s.	days after sight	見票後…日付款
DV	dividends	紅利，股息
E		
ea.	each	每、各
EC	error correction	錯誤更正
e.e., E.E.	error excepted	錯誤除外
EEC	The European Economic Community	歐洲共同市場（經濟體）
e.g.	exempli gratia (for example)	例如
EIB	Export-Import Bank	進出口銀行
enc., encl.	enclosure	附件
EOD	end of discussion	討論完畢
E&O.E.	errors and omissions excepted	錯誤或遺漏不在此限
E.O.M.	end of month	月底
ER	error	錯誤
esp.	especially	尤其
ETA	estimated time of arrival	預定到達日
etc.	et cetera (and so forth)	…等等
ex	out of, from, without	無，出自，除出

ex.	example, executive, exchange, extract	例子，主管，外匯交換，摘要
Excl.	exclude, exclusive	除外
exp	export	出口
EZ	easy	簡單
F		
F2F, FTF	face to face	面對面
f	foot, franc	呎，法郎
f.a.a	free of all average	全損才賠
FAQ	frequently asked questions	常問的問題
f.a.s.	Free alongside Ship	船邊交貨價
F.B.E.	Foreign Bill of Exchange	國外匯票
FCFS	first come, first served	先到先服務
f.c.l.	full container load	整貨櫃裝滿
f.d.	free discharge	卸貨船方不負責
F.D.	foreign department, foreign division	國外部
f.i.	free in	裝貨船方不負責
FITB	fill in the blank	填空
f/o	for account of	入…帳
F/O	for orders	應付訂單
F.O.B.	Free on Board, Free on Board of Plane	船上交貨價，離岸價，飛機上交貨價
f.o.c.	free of charge	免費
f.o.s.	free of Steamer	輪船上交貨價
f.o.t.	Free of Truck	卡車上交貨價
f.p.a.	Free of Particular Average	單獨海損不賠
F/P	floating policy	流動保單
F.X.	foreign exchange	外匯
FY	fiscal year	會計年度
FYI	for your information	供您參考
FYR	for your reference	供您參考
G		
g.	good, goods, gram	佳，貨物，公克
G/A	General Average	共同海損
GA	go ahead	前進
GATT	General Agreement on Tariffs and Trade	關稅貿易總協定
GDP	gross domestic product	國內生產毛額

GIWIST	gee, I wish I said that	其實我也想那麼說
g.m.	gram	公克
GMAT	Graduate Management Admission Test	管理學院研究生入學測驗
GNP	gross national product	國民生產毛額
govt.	government	政府
GP	gross profit	毛利
GPO	General Post Office	郵政總局
g.s.w.	gross shipping weight	裝輸總重量
G.W.	gross weight	毛重
H		
H.O.	Head Office	總公司
h.p.	horse power	馬力
HQ	high quality, head quarter	高品質，總部
hrs	hours	小時
I		
IAC	in any case	在任何情況中
IAE	in any event	在任何事件中
IBRD	International Bank for Reconstruction and Development	國際建設開發銀行
IC	I see	我明白
I/C	inward collection	對內催收
ICC	International Chamber of Commerce	國際商會
ICQ	I see you	我找你
i.e.	id est (that is)	即是，就是
IMCO	in my considered opinion	我仔細考慮過
IMF	International Monetary Fund	國際貨幣基金
IMHO	in my humble opinion	依我的淺見
IMMED/IMDTLY	immediately	立刻
IMO	in my opinion	依我之見
imp.	import	入口
IN	interest	利息
Inc.	incorporated	法人，股份有限公司
incl.	include	包括
Incoterm	International Terms	國際商業用語
info	information	資料

INQRY	inquiry	詢問
inst.	instant (this month)	本月
Insur.	insurance	保險
int.	interest	利息
inv.	invoice	發票
IOU	I owe you	借據
IOW	in other words	換句話說
I/R	inward remittance	內匯
ISIC	International Standard Industrial Classification	國際工業標準分類
ISO	International Standardization Organization	國際標準化組織
IT	item	項目
ITMT	in the meantime	那麼，另一方面
IWBNI	it would be nice if	如果…就好了
J		
J/A	joint account	聯合帳戶
JAM	just a minute	等一下
K		
k.	carat	克拉
kg.	keg, kilogram	小桶，公斤
K.W.	Kilo Watt	千瓦
L		
£	pound or pounds sterling	英鎊
L8R	later	等一下
L/A	letter of account	帳目通知書
lb	libra (pound or pounds in weight)	磅
lbs.	pounds	磅
L/C	letter of credit	信用狀
L/G	letter of guarantee	保證函
L/H	letter of hypothecation	質權書
L/I	letter of indemnity	賠償保證書
LIFO	last-in first-out	後進先出（法）
l.t.	long ton	長噸
L/T	letter Telegram	書信電報
Ltd.	Limited	有限公司
LTR	letter	信件，書信

M		
M	Monsieur	先生
Max	maximum	最大限度
m/d	month after date	出票後…月付款
mdse.	merchandise	商品
memo	memorandum	備忘錄、便條紙
mfg.	manufacturing	製造的，製造業
mfr.	manufacturer	製造商
mg.	milligram	毫克
MGMT	management	管理
M.I.	marine insurance	海上保險
MIN, min	minimum	最少，最低限度
mins	minutes	分鐘
M.I.P.	marine insurance policy	海上保險單
misc	miscellaneous	雜項
mk	mark	嘜頭
mkt	market	市場
M/L	more or less	約計
Mme	Madame	夫人
MO	Money Order	匯款單
MR	Mate' Receipt	大副收據
m/s	months after sight	見票後…月付款
m.s.	mail steamer, motor ship	郵船，柴油船
MSG	message	信息，電文
M.T.	merit ton, mail transfer, Motor Tanker	公噸，信函，柴油機油輪
m.v.	motor vessel	柴油機船
N		
N.B.	Nota bene (take notice)	注意
NBD	no big deal	沒什麼了不起
NEC	necessary	必要的
n/f	no fund	無經費
NL	night letter	夜間發電報
NO.	number	號碼，編號
n/p	non-payment	拒付
NP	net profit	淨利
NQA	no questions asked	沒有問題了
N.W.	Net Weight	淨重

O		
O/	to the order of	由…指示
OBO	or best offer	或更好的選擇
OBTW	oh, by the way	可是，再說
O/C	outward collection	對外催收
O/D	overdraft, on demand	透支，要求即付款
OIC	oh, I see	嗯，我明白了；原來如此
O.K.	all correct, approved, alright	無誤，同意，沒問題
O/No.	order number	訂單編號
o.p.	open policy	預約保單
O/R	outward remittance	對外匯款
ORD	ordinary telegram	尋常電報
o/s	on sale, out of stock	廉售，無庫存
O/S	old style	老款式
o.t.	old term	舊條件
OTOH	on the other hand	另一方面
OTP	on the phone	在電話中
OTTH	on the third hand	第三方面
oz	ounce	盎司
P		
P/A, p/a	particular average	單獨海損
p.a.	per annum (per year)	每年
P.A.Y.E.	pay as you enter	見貨即付
pc(s)	piece(s)	件
P.C.	percent	百分比
Pcks	packages	包裝，包裹
Pd.	paid	已付
per pro	per procuration (by power of authority)	代理
p.f.	per forma invoice	估價發票
p.g.t.	per gross ton	每毛噸
p.h.	per hour	每小時
P&L	profit and loss	損益
p.m.	post meridiem (past noon)	下午
P.M.O.	postal money order	郵政匯票
P/N	part number, promissory note	零件號碼，本票（期票）
P/O	payment order	付款訂單
p.o.d.	payment on delivery	交貨時付款

P.O.B.	post office box	郵政信箱
POV	point of view	觀點，想法
pr	power of attorney	委任狀
P/R	parcel receipt	郵包收據
pro forma	for form's sake	形式的
PS.	postscript	後記
PT.	payment	付款
P.T.O.	please turn over	請轉下頁
PTL	private tieline service	電報專線業務
Q		
qlty	quality	品質
qtr	quarter	四分之一，一季
qty	quantity	數量
quotn	quotation	報價
qy	quay	碼頭
R		
R	are	是…
RE	regarding	有關
R/E	Rate of Exchange	匯率
recd	received	收訖
recpt	receipt	收據
ref.	reference	參考
remit, rm	remittance	匯款
RGDS	regards	請安
r.m.	ready money, ready-made	備用金，現成的
Ro	remittance order	匯款訂貨
R.P.	reply paid	郵費或電報費預付
RSN	real soon now	不久，立刻
R.S.V.P., RSVP	please reply	請回音
RT.	returned unpaid item	未付款退貨
rt.	rate	率
S		
S.A.	Statement of Account	帳單
s.a.	subject to approval	以贊成為前提
sc.	service charge	服務費
sd.	sundries	雜質
S/D	sea damage	海水損害
secy	securities	證券、擔保、安全

sig.	secretary	祕書
S/N	signature	簽名
S.O.	shipping note	裝船通知
SRCC	strike, riot, civil commotions	罷工，暴動，內亂險
s/s, ss, s.s.	steamship	輪船
s.t.	short ton	短噸
st.	street	街
SUP	what's up	出了什麼事？怎麼樣？
s.v.	sailing vessel	帆船
SVP	please	請
T		
T/A	telegraphic address	電報掛號
TC	collated telegram	校對電報
tgm	telegram	電報
THX	thanks	謝謝
TIA	thanks in advance	先道謝了
T.L.O.	total loss only	只擔保全損
T.M.O.	telegraphic money order	電報匯款
Tn.	ton or tons	噸
T.P.N.D.	Theft, Pilferage and Non-Delivery	未送達險全險
T.R., T/R	trust receipt	信託收
T.T., T/T	telegraphic transfer (wire transfer)	電匯
TTYL	talk to you later	稍後再說
U		
4U	for you	給你
U	you	你
UL, U/L	upload	上傳
uos	unless otherwise specified	除非另有約定
UR	you are, your	你是…，你的
u/w	underwriter	保險業者
V		
val	value	價值
VAT	value added tax	增值稅
vd	value date	價值日期
VIP	very important person	重要人物，貴賓

viz	videlicet (namely)	即是
voy.	voyage	航次
W		
w.a.	with average	水漬險（單獨海損賠）
W.A.R.	with all risk	擔保一切險
WB	welcome back	歡迎回來
W/B	way bill, warehouse book	貨運單，倉庫簿
WBR	with best regards	順致問候
WBW	with best wishes	致以最好的祝福
wgt	weight	重量
whf	wharf	碼頭
W/M	weight or measurement	重量或體積
w.p.a.	with particular average	單獨海損賠償
W.R.	War Risk	兵險
WRT	with respect to, with regards to	關於，（至於）談到
WYSIWYG	what you see is what you get	你所見的就是你所得的
X		
x	ex (out of, without), exclusive	由，除外
x.c.	ex coupon	無息票
x.d.	ex dividend	紅利（或股息）未付
x.i.	ex interest	不含利息
Y		
yd.	yard (measurement)	碼
YMMV	your mileage may vary	或許意見和你不同

III. 部門名稱中英對照表

會計部	Accounting Department
行政管理部	Administration Department
廣告課	Advertising Division
稽核室	Auditing Office
分公司	Branch Office
顧客服務部	Customer Relations/Service Department
設計部	Design Division
國內銷售	Domestic Sales
出口部	Export Department
財務部	Finance Department
總務部	General Affairs Department
總公司	Head Office
人力資源處	Human Resource Department
資訊系統部	Information System Department
後勤支援部	Logistics Department
維修部	Maintenance Section
行銷部	Marketing Department
總裁／總經理辦公室	Office of the President
祕書處	Office of the Secretary
營運／廠務部	Operations Department
國外銷售部	Overseas Sales Division
零件部	Parts Section
專利課	Patent Section
人事課	Personnel Section
企劃課	Planning Section
資材課	Procurement Section
生產部	Production Department
公關部	Public Relations/Publicity Section
採購部	Purchasing/Procurement Department
品管課	Quality Assurance/Quality Control Section
研發部	Research and Development Department
業務部	Sales Department
保全	Security Division
出口部	Shipping Department
倉庫課	Storage Section
運輸課	Transport Section

Ⅳ. 職務名稱中英對照表

執行長	CEO (Chief Executive Officer)
財務長	CFO (Chief Financial Officer, Controller)
營運長	COO (Chief Operation Officer)
會計	Accountant
顧問／高級顧問	Advisor/Senior Advisor
助理	Assistant
助理工程師	Assistant Engineer
副理	Assistant Manager
副理／助理部們經理	Assistant Department Manager
總經理特助	Assistant to President
稽核	Auditor
總裁	Chairman
事務員	Clerk
顧問	Consultant
部門經理	Department Manager
副課長／副科長	Deputy (Section) Manager
副理	Deputy Department Manager
處長	Director
董事	Director of Board
工程師	Engineer
行政顧問	Executive Advisor
總經理	General Manager
名譽主席	Honorary Chairman
經理	Manager
協理／執行董事	Managing Director
作業員	Operator
廠長	Plant manager
董事長	President
首席工程師	Principal Engineer
祕書	Secretary
課長／科長／主任	Section Manager
資深工程師	Senior Engineer
資深執行董事／協理	Senior Managing Director
首席副總裁	Senior Vice President
特助	Special Assistant
專員	Specialist

職員	Staff
主管／主任	Supervisor
系統工程師	System Engineer
組長／領班	Team Leader
技術員	Technician
副主席／副董事長	Vice Chairman
副總裁／副董事長	Vice President

(1) 以前稱「執行長」為 Managing Director，現在一般都稱 CEO
由於各國和各個公司的職務編制有不同的職級制度，此表的職務名稱僅供參考。

V．政府機關英文名稱

1. 總統府【Office of the President ,Republic of China】

- 中央研究院：Academia Sinica

2. 立法院【The Legislative Yuan of R.O.C】

- 內政及民族委員會：Home and Nations Committee
- 外交及僑務委員會：Foreign and Overseas Chinese Affairs Committee
- 科技及資訊委員會：Sci-Tech and Information Committee
- 國防委員會：National Defense Committee
- 經濟及能源委員會：Economics and Energy Committee
- 財政委員會：Finance Committee
- 預算及決算委員會：Budget and Final Accounts
- 教育及文化委員會：Education and Culture Committee
- 交通委員會：Transportation Committee
- 司法委員會：Judiciary Committee
- 法制委員會：Organic Laws and Statutes Committee
- 衛生環境及社會福利委員會：Sanitation and Environment, as well as Social Welfare Committee
- 程序委員會：Procedure Committee
- 紀律委員會：Discipline Committee
- 修憲委員會：Constitutional Amendment Committee
- 經費稽核委員會：Expenditure Examination Committee

3. 監察院【The Control Yuan of R.O.C】

- 審計部：Ministry of Audit R.O.C

4. 考試院【The Examination Yuan of R.O.C】

- 考選部：Ministry of Examination
- 銓敘部：Ministry of Civil Service

5. 司法院【The Judicial Yuan of The Republic Of China】

- 公務員懲戒委員會：COMMISSION ON THE DISCIPLINARY SANCTIONS OF FUNCTIONARIES
- 最高法院：THE SUPREME COURT
- 最高行政法院：THE SUPREME ADMINISTRATIVE COURT
- 臺北高等行政法院：THE TAIPEI HIGH ADMINISTRATIVE COURT
- 臺灣高等法院：THE TAIWAN HIGH COURT
- 臺灣高等法院臺中分院：THE TAIWAN HIGH COURT TAICHUNG BRANCH COURT
- 臺灣台北地方法院：THE TAIWAN TAIPEI DISTRICT COURT

6. 行政院【The executive Yuan of the Republic of China】

- 內政部：Ministry of the Interior
- 外交部：Ministry of Foreign Affairs
- 財政部：Ministry of Finance
- 教育部：Ministry of Education
- 法務部：Ministry of Justice
- 經濟部：Ministry of Economic Affairs
- 交通部：Ministry of Transportation and Communications
- 蒙藏委員會：Mongolian & Tibetan Affairs Commission
- 僑務委員會：Overseas Chinese Affairs Commission
- 主　計　處：Brief Introduction to Directorate-General of Budget, Accounting and Statistics

- 人事行政局：Central Personnel Administration
- 新聞局：The Government Information Office
- 衛生署：Department of Health
- 環境保護署：Environmental Protection Administration
- 故宮博物院：National Palace Museum
- 大陸委員會：Mainland Affairs Council
- 經濟建設委員會：Council For Economic Planning And Development
- 青年輔導委員會：National Youth Commision
- 國軍退除役官兵輔導委員會：Veterans Affairs Commission
- 原子能委員會：Atomic Energy Council
- 國家科學委員會：National Science Council
- 農業委員會：Council of Agriculture
- 文化建設委員會：Council for Cultural Affairs
- 公平交易委員會：Fair Trade Commission
- 公共工程委員會：Public Construction Commission
- 原住民族委員會：Council of Indigenous Peoples
- 體育委員會：National Council on Physical Fitness and Sports
- 中央銀行：The Central Bank of China
- 消費者保護委員會：Consumer Protection Commission
- 飛航安全委員會：Aviation Safety Council
- 海岸巡防署：Coast Guard Administration
- 研究考核委員會：Research, Development, and Evaluation Commission
- 中央選舉委員會：Central Election Commission
- 客家委員會：Council for Hakka Affairs
- 勞工委員會：Council for Labor Affairs
- 北美事務協調委員會：Coordination Council for North American Affairs

Ⅵ. 銀行中英文名稱對照表

國內銀行：

1	台灣銀行	BOT, Bank of Taiwan
2	永豐銀行	Bank SinoPac
3	日盛銀行	JihSun Bank
4	高雄銀行	BOK, Bank of Kaohsiung
5	元大銀行	Yuanta Bank
6	中國信託	CTCB, Chinatrust Commercial Bank
7	匯豐銀行	HSBC Direct
8	玉山銀行	E. Sun Commercial Bank
9	萬泰銀行	CSMB, Cosmos Bank
10	彰化銀行	CHCB, Chang Hwa Commercial Bank
11	安泰銀行	Entie Commercial Bank
12	第一銀行	FCB, First Commercial Bank
13	華南銀行	HNCB, Hua Nan Commercial Bank
14	兆豐銀行	Mega International Commercial Bank
15	土地銀行	LBOT, Land Bank of Taiwan
16	遠東國際商業銀行	FEIB, Far Eastern International Bank
17	大眾銀行	TACB, Ta Chon Commercial Bank
18	上海銀行	SCSB, Shanghai Commercial Savings Bank
19	台新銀行	TSIB, Taishin Business Bank
20	國泰世華銀行	UWCB, United World Chinese Commercial Bank
21	合作金庫	TCB, Taiwan Cooperative Bank
22	台灣企銀	TBB, Taiwan Business Bank
23	富邦銀行	Fubon Financial Bank
24	交通銀行	Bank of Communications
25	華僑銀行	Overseas Chinese Bank
26	中國輸出入銀行	The Export-Import Bank of the R.O.C.
27	聯邦商業銀行	Union Bank of Taiwan
28	華泰銀行	Hwaiti Commercial Bank

國外銀行：

1	美商花旗銀行	CITIBANK N.A
2	美國商業銀行	Bank of America NT&SA
3	美國運通銀行	American Express Bank Ltd.

4	美商大通銀行	Chase Manhattan Bank
5	美國紐約銀行	Bank of New York
6	美商波士頓銀行	Bank Boston ,N.A
7	美商信孚銀行	Bankers Trust Company
8	美商加州聯合銀行	Union Bank of California N.A
9	美商費城國民銀行	Core States Bank N.A
10	美商明尼蘇達西北銀行	Norwest Bank Minnesota
11	美商道富銀行	State Street Bank & Trust Co.
12	美商眾國銀行	Nations Bank N.A.
13	美商夏威夷銀行	Bank of Hawaii
14	美國利寶銀行	Republic National Bank of NewYork
15	美國加州廣東銀行	Bank of Canton of California
16	美商摩根銀行	Moman Guaranty Trust Companyof New York
17	美商芝加哥第一國民銀行	The First National Bank of Chicago
18	加拿大商多倫多道明銀行	Toronto-Dominion Bank
19	加拿大皇家銀行	Royal Bank of Canada
20	加拿大商豐業銀行	The Bank of Nova Scotia
21	加拿大帝國商業銀行	Canadian Imperial Bank of Commerce
22	加拿大國家銀行	National Bank of Canada
23	加拿大蒙特利爾銀行	Bank of Montreal
24	日商第一商業銀行	Dai-Ichi Kangyo Bank
25	日商東京三菱銀行	Bank of Tokyo Mitsubishi Ltd.
26	日商東海銀行	The Tokai Bank, Ltd.
27	日商富士銀行	The Fuji Bank Ltd.
28	日商三和銀行	The Sanwa Bank, Ltd.
29	日商櫻花銀行	Sakura Bank
30	日商旭日銀行	The Asahi Bank Ltd.
31	新加坡國際銀行	International Bank of Singapore
32	新加坡發展銀行	The Development Bank of Singapore
33	新加坡大華銀行	United Overseas Bank Limited
34	新加坡商新加坡華僑銀行	Oversea-Chinese Banking Corporation Ltd.
35	新加坡達利銀行	Tat Lee Bank Ltd.
36	香港上海匯豐銀行	The Hongkong & Shanghai Banking Corp. Ltd.
37	香港東亞銀行	The Bank of East Asia, Ltd.
38	香港道亨銀行	Dao Heng Bank
39	香港亞洲商業銀行	Asia Commercial Bank Ltd.
40	泰國盤谷銀行	Bangkok Bank Public Company Ltd.
41	菲律賓首都銀行	Metropolitan Bank & Trust Company

42	菲商菲律賓國家銀行	Philippine National Bank
43	菲商菲律賓土地銀行	Land Bank of Philippines
44	菲律賓國際商業銀行	Philippine Commercial International Bank
45	澳洲國民銀行	National Australia Bank
46	澳商澳洲紐西蘭銀行	Australia & New Zealand Bank
47	英商渣打銀行	Standard Chartered Bank
48	英國米特蘭銀行	Midland Bank Pic.
49	英商國民西敏銀行	National Westminster Bank Pie
50	法國興業銀行	Societe Generale
51	法商百利銀行	Banque Paribas
52	法國國家巴黎銀行	Banque National De Paris
53	法國東方匯理銀行	Credit Agricole Indosuez
54	德商德意志銀行	Deutsche Bank AG
55	德商德利銀行	Dresdner Bank A.G.
56	德商西德意志州銀行	Westdeutsche Landesbank
57	德商德國商業銀行	Commerzbank A.G.
58	荷商荷蘭銀行	ABN AMRO Bank N.V.
59	荷商荷興銀行	ING Bank N.V.
60	荷商荷蘭合作銀行	Rabobank Nederland
61	南非標旗銀行	The Standard Bank of South Africa Ltd.
62	南非萊利銀行	Nedcor Bank Limited
63	瑞士商瑞士聯合銀行	Union Bank of Switzerland
64	瑞士商信貸銀行	Credit Suisse
65	瑞士商瑞士銀行	Swisse Bank Corporation
66	比利時商信貸銀行	Kredietbank N.V.,
67	比利時商通用銀行	Generale Bank
68	瑞典商瑞典商業銀行	Syenska Handelsbanken
69	西班牙國際商業銀行	Banco Santander
70	義大利商意大利商業銀行	Banca Commerciale Italiana
71	義大利商嘉利堡銀行	Cariplo S.P.A.

VII. 國際貨幣與代號

常見國家外幣與唸法：

美洲

國家	貨幣名稱	貨幣符號	念法	代號
美國	美元	US$	U.S. Dollar	USD
加拿大	加拿大元	CAN$	Canadian Dollar	CAD

亞洲

國家	貨幣名稱	貨幣符號	念法	代號
台灣	新台幣	NT$	New Taiwan Dollar	NTD
中國	人民幣	RMB ¥	Renminbi Yuan	CNY
香港	港幣	HK$	HongKong Dollar	HKD
日本	日圓	¥	Japanese Yen	JPY
韓國	韓元	₩	Korean Won	KRW
澳門	澳門元	P	Macanese pataca	MOP
越南	越南盾	₫	Vietnamese Dong	VND
泰國	泰銖	฿	Thai baht	THB
菲律賓	菲律賓披索	₱	Philippine peso	PHP
新加坡	新加坡元	S$	Singapore dollar	SGD
柬埔寨	瑞爾	៛	Cambodian riel	KHR
馬來西亞	令吉	RM	Malaysian ringgit	MYR
緬甸	緬甸元	K	Kyat	MMK
印度	印度盧比	Rs / ₹	Indian rupee	INR
印尼	印尼盾	Rs	Indonesian rupiah	IDR

大洋洲

國家	貨幣名稱	貨幣符號	念法	代號
澳洲	澳幣	A$	Australian Dollar	AUD
紐西蘭	紐西蘭元	NZ$	New Zealand dollar	NZD

歐洲

歐洲聯盟裡的成員，像是荷蘭、法國、西班牙、德國、義大利等成員國都是使用歐元；歐盟以外的國家大多數是使用自己國家的貨幣，只有少部分國家需要歐盟經濟保護，還是持續使用歐元。

國家	貨幣名稱	貨幣符號	念法	代號
英國	英鎊	£	Pound, Sterling	GBP = Great Britain Pound
歐洲聯盟	歐元	€	Euro	EUR
捷克	捷克克朗	K	Czech koruna	CZK
瑞士	瑞士法郎	Fr	Swiss franc	CHF

英國貨幣由英鎊 (pound) 和便士 (pence) 組成，1 英鎊等於 100 便士。與美國錢幣相似，但美國稱紙幣多稱為 bills；而英國稱 notes。

非洲

國家	貨幣名稱	貨幣符號	念法	代號
埃及	埃及鎊	£ / م.ج	Egyptian pound	EGP

▶ Text 練習題解答

Chapter 01

(b) 1. Which one is not part of AIDA?
(a) Desire (b) Attraction (c) Interest

(c) 2. Which is one of the eight Cs of good writing
(a) Coaching (b) Compelling (c) Courtesy

(a) 3. Which sentence is better?
(a) Our next meeting will be held at 9 am tomorrow.
(b) Please be advised that our next meeting will be held at 9 am tomorrow.

(a) 4. What is KISS?
(a)keep it as simple as possible (b)keep it as special as possible
(c)keep it as speedy as possible

Chapter 02

選擇題：

(d) 1. Which is not an appropriate way of writing an attention line?
(a) Attention: Ben Lin (b) Attn. Ben Lin
(c) Attention Ben Lin (d) all are acceptable.

(b) 2. Which is not an appropriate salutation for a business letter?
(a) Dear Mr. Jorden (b) Hello (c) Ladies and Gentlemen

(b) 3. Which is a correct format for the date in a business letter
(a) 21-09-21 (b) September 21, 2021 (c) 9/21/2021

(c) 4. Which is correct?
(a) Yours Truly (b) yours truly (c) Yours truly

問答題：

1. 請標示書信中的欄位名稱

① Sender's address
② Date
③ Recipient's address
④ Salutation
⑤ Subject
⑥ Body of letter
⑦ Closing/complimentary close
⑧ Signature
⑨ Name
⑩ Title
⑪ Identification initials
⑫ Carbon copy
⑬ Enclosure

2. 請就第一題書寫信封

Ans:

Alex Chen
Taiwan Bear International
3F, No. 11, Park Avenue II
Science-Based Industrial Park
Hsin-Chu 30075
Taiwan

Donald Liu
Research and Development Dept.
SemiTech Group
309 Ditmas Ave
Brooklyn, NY 11218
USA

3. 請就第一題書寫傳真的封面

Ans:

Taiwan Bear International
3F, No. 11, Park Avenue II
Science-Based Industrial Park
Hsin-Chu 30075, Taiwan
Tel: 886-03-5798888
Fax: 886-035978891

FAX

Company : Semi Tech Group
Attn : Mr. Donald Liu
Fax No. : (718)656-7000

Date : Aug. 18, 2021
Pages : 3 pages

4. 請就下列情境寫一份 memo：人力資源處長將告訴公司所有員工於今年 11 月底前，所有員工需於電腦問卷上完成「員工滿意度」調查。

Ans:

Taiwan Bear International

MEMO

To: All Employees
From: HR Director
Cc: President
Date: Nov. 01, 2021
Re: Employee Satisfaction Survey

Dear All,

This is an announcement to all employees who are requested to finish "Employee Satisfaction Survey" on internet by the end of November.

Your correspondence will be appreciated.

Thanks

Chapter 03

1. 將下列的句子翻成英文

A. We want to tell you about a new type of furniture that is selling like hot cakes in the USA.
B. Our products are of the finest materials and the highest techniques and are second to none in design and reliability.
C. Please pay your best attention to the new specialties which made a great sensation at this year's International Trade Fair.
D. You will be enthusiastic about the enclosed catalogs showing our complete range of new toys for this Christmas season and New Year.
E. You will be interested to know that we have just introduced our new micro oven.
F. We are enclosing some sales literature in which you will see what an enthusiastic reception our new product has received all over the world.
G. We got your name from the Trade Sources and learned that you are one of the leading importers of computer products in your country.
H. As a well-experienced exporter, we have been working in the field of computer products since 1980 and also have very good relationship with local manufacturers.

I. We are one of the biggest manufacturers of toys and have the confidence to satisfy our customers in every respects.
J. Through the recommendation of your Chamber of Commerce, we are writing you this letter in order to know if you are interested in importing the computer products.

2. 將下列中文書信翻譯成英文

We are glad to know from the Hannover Fair that you are the leading importer of computer products in Germany, and would like to take this opportunity to introduce you a high technical product which was just developed by us this Year.

With many years' experience in R&D and many professional engineers, we can constantly offer the newest products and create the best business opportunity for the customers. Enclosed are our relevant catalog and quotation for reference.

Thank you for your interest and hope to set up business relationship with you very soon.

3. 請自由發揮寫一封完整的鞋子的推銷信，註明我們的工廠在大陸可以提供價廉物美的商品，消息來源設定為從對方國家的商務部。

We are glad to know from your Chamber of Commerce that you are a leading importer of shoes, and would like to take this chance to introduce you our products.

From the enclosed catalog and price list, you will find we can offer the most fashion style products at the lowest prices as we have our own factory in China.

Please take the above into consideration and don't hesitate to let us know of your interest soon.

4. 寫一封參展【台北文具春季展】後的推銷信，對換過名片的潛在客戶去函告知寄給對方我們的商品型錄。

Dear Mr. White,

It was our pleasure to have met you at the Taipei Stationery Spring Fair. We are pleased to have a favorable discussion with you.

We believe you will be interested in our new products. Therefore, I send you our latest catalog in which you could find our complete products with details, such as function and specification, etc.

Please don't hesitate to contact us if you need any further information.

Best regards,

5. 請寫一封推銷信，信中說明你目前在一家航空公司工作，你是從維他命出口的網頁上得知 Jolie Boutique。說明此航空公司已成立超過 8 年，你的公司有良好的商譽，並且送貨迅速而值得信賴，從未掉過包裹。所以想提供對方一個優惠的運送價格。因為公司與其他運輸業者有良好的商業關係，因此可以提供更低廉的價格。

Dear Ms. Lee

I came across your boutique on a Vitamin Exporters website, and I believe I can give you a more competitive price for your shipments than you are currently receiving.

Please allow me to tell you a bit about Air Freight Inc. We have been in business for more than eight years. In that time, we have built a reputation for reliability and speedy service. Our customers trust us because we have never lost a package.

We are also known as one of the most competitively priced shipping agencies, as our unique business relationships with other shippers allow us to combine freight to get the best prices for our customers.

I hope you will visit our website at www.airfreight.tw for more information about rates and services. Please also feel free to call 886-2-22879966 at any time to discuss how we can benefit your business.

Sincerely yours,

Chapter 04

1. 將下列的句子翻成英文

A. Please send a catalog of your latest products.
B. Kindly send us without charge some samples of your products.
C. We should appreciate full particulars of your newly-developed product.
D. We hear that you have put an electric car on the market and should be glad to have full details.
E. We are interested in your products. Do you have any brochures which you could send us?
F. Please send us your catalog and quotation for evaluation.

2. 將下列中文書信翻譯成英文

We know your name from the Far East Trade Office and are interested in your product.

We are the largest importer of sport shoes in US. We have over 500 sales and very good sales channels, and also have very good relationship with buyers. So, please email us your best quotation showing the price scale.

Thank you for your assistance and awaiting your reply.

3. 請寫封詢價信函（感謝賣方寄來的目錄及公司光碟，研讀後對 MRI 系列極有興趣，請賣方報最好價格，並請寄一套樣品來）。

Dear Sirs,

Thank you for your cooperation in sending us your latest catalogues which we just received. After going through them, we are very interested in your MRI-series. Please kindly quote your best prices and send us a set of samples further.

We look forward to a favorable business with you. Your prompt cooperation about the above will be highly appreciated.

Best regards,

4. 寫一封詢問函，請求賣方提供最新的商品型錄與報價。（產品自定）

Dear Sirs,

We saw your advertisement in Asian Magazine and are interested in importing your products.

With many years' experience in selling the same products, we have already set up very good sales channels here, and have the confidence to promote your products well in our market.

Please send us your catalogs and quotation soon.

Best regards,

5. 寫一封詢價信函，表示在雜誌「The Houseware」上看到賣方的商品廣告，我們對日式茶壺和碟子有興趣。請賣方報價 CIF LA 及價格折扣、出貨日與付款方式，並請提供一套樣品，以供評估。

Dear Sirs,

We have seen your advertisement in "The Houseware" and are interested in your Japanese teapot and dish.

Please kindly quote your best prices based on CIF L.A., stating your earliest delivery date, your terms of payment and discounts. Also, please send samples for evaluation.

We look forward to a long and happy business relationship with you.

Best regards,

Chapter 05

1. 將下列的句子翻成英文

A. We offer these goods on very favorable terms at 30 days under D/A.
B. We offer you this product at prices ranging from US$ 30 to US$ 33.
C. For quantities of more than 1000 pieces, we offer a discount of 5% on the list prices.
D. Owing to the slump in commodity prices, we offer you these goods at 15% below the market price.
E. We are pleased to offer you firm the following goods subject to your reply reaching us by end of April.
F. If you miss this opportunity, you may not be able to obtain the item even at a higher price.
G. As you will know, the prices have risen sharply during the last few weeks. So this is a very moderate increase, and we hope that you will take advantage of our offer.
H. We advise you to accept this offer without loss of time.
I. Thank you for your inquiry of 9/28 and are glad to quote as follows:
J. The above offer is the net price without our commission.

2. 將下列中文書信翻譯成英文

Dear Sirs,

Thank you for your e-mail of 3/2 with interest in our sweaters. As requested, we are glad to quote as below:

Price: US$35.60/dz FOB Taiwan
Delivery: in 30 days after receiving the order
Payment: by irrevocable and confirmed L/C at sight
Min. Q'ty: 1000 dz
Validity: 30 days

If you need further information, please advise without hesitation. We are waiting for your order soon.

Best regards,

3. 在信件中告訴客戶，報價單如附件，請客戶參考，並告知意見或下訂單。

Dear Sirs,

Thank you for your e-mail of June 29. As requested, we enclose our price list for the items in which you are interested. We hope you are satisfied with our offer and can tell us your comments or place an order soon.

We look forward to the opportunity of serving you further.

Best regards,

4. 寫一封在 e-mail 中直接報價給新客戶的信函，再次介紹公司，強調產品的優點，希望客戶能有興趣並儘早下單。

Dear Sirs,

Thank you for your e-mail of May 5th inquiring for our 141000 series. As requested, we quote our best FOB TAIWAN price as follows:

Item No.	FOB TAIWAN	Packing/Meas./N.W./G.W.	
141001	US$2.09/pc	480pc /2.08'	13/14 kgs
141002	US$2.15/pc	480pc /1.75'	12/13 kgs
141003	US$3.45/pc	720pc /3.19'	15/17 kgs

* The above prices are based on one logo, 4-color printing.
* Delivery: within 30 days after receipt of your L/C or T/T.
* Payment: by irrevocable L/C at sight in our favor or T/T before shipment.
* Samples cost: US$200 as plate charge and sample charge, which will be deducted upon order confirmation.
* Minimum: 1,000 pcs PER ITEM.

We are a professional supplier in the industry of promotion products in Taiwan. We provide the wide selection of high quality products, on time and in accordance with specifications. Regarding the series you requested, we have sold very well in the world. They have nicer outlooks and more functions than the old models. We hope that you also agree that the new models are more favorable and competitive, and can confirm order as soon as possible.

Best regards,

5. 感謝客戶參展時的訪問，附上當時所談商品的報價單，希望客戶及早訂購。

Dear Sirs,

It was our pleasure to meet you at the Taipei stationery fair. Please refer to the products we discussed at the fair, we enclose our former quotation for your further reference, which we have listed the packing and the terms in detail.

We believe a trial order will convince you of the superior quality with competitive prices of our products. Your prompt action by confirming an order return will be very favorable.

We look forward to hearing from you soon.

Best regards,

6. 感謝客戶來函表示對我方產品的興趣即索取報價。但是由於我方商品項目繁多，請客戶先於公司型錄上挑選有興趣的產品，我方再進一步報價。

Dear Sirs,

Thanks for your email yesterday with interest in our products. However, due to varieties of our goods, we could not give you the quotation very soon. If you can select interested items from our catalog and advise, we can make quotation as fast as we can.

Best regards,

7. 漲價通知，發函給舊客戶。由於工資、物料的上漲，以及台幣的升值，我方價格將於 9 月 1 日起全面調漲 3%。

Dear Sirs,

We advise here that all of our prices will be increased 3% effective from September 1st, 20XX, due to the increase of accessories and the appreciation of NT dollars. We are sorry to have this increase, but we have no choice. Please kindly understand.

Your prompt order confirmation before September 1st will remain the old prices. Please kindly take it into your consideration and confirm orders at the soonest.

Best regards,

8. 告訴客戶無法報價，因為 AA 系列已經停產，目前的新商品是 BB 系列，我們可以報 BB 系列的價格，希望客戶可以接受。

Dear Sirs,

Thank you for your inquiry for a quotation of our AA-series. We are glad to inform you that it's no longer in production and the new substitute is BB-series. Please refer to the pictures and quotation, we hope that you agree that this series is really an excellent substitute and can consider placing an trial order return. So, we look forward to your comments soon.

Best regards,

Chapter 06

1. 將下列的句子翻成英文

A. In view of the prevailing prices in this market, your quotation is a little expensive.
B. A comparison of your offer with that of our regular supplier shows that their quotation is more favorable.
C. Thank you for answering our inquiry. Unfortunately, your prices are too high for our market.
D. It would not be possible for us to purchase the quantity at the price you offer.
E. This order is quite large, and we would request that you make a quantity discount.
F. We can oly consider placing an order if you can give a price reduction of 10%.
G. We will work harder to secure orders from clients if you assist us with lower prices.
H. As business has been rather slack, we would ask you to alter your payment terms.
I. If you can meet the terms we require, we are prepared to place an order with you.
J. In view of our difficulty in payment, we ask you to extend the sight of your draft to 90 days.
K. In view of the long-term business relations with your company, we accept the terms you specified for this order.
L. Since we expect to enter your market, we will cut our margin of profit to give you the benefit of a 4% reduction.
M. We agree to make a reduction in price if this will help you to develop your market for our products.
N. We are prepared to give you an additional allowance of 5% on an order of 1000 units or over.
O. Our target price is US$10/pc, please advise if you can accept.
P. Your price is 5% higher than those of the competitors.

2. 將下列中文書信翻譯成英文

Dear Sirs,

Thank you for your email of March 2nd and quotation for your latest sweater in catalog, but we found your price too high.

As our order quantity is 10,000 dozens, please re-quote a competitive price to us. Our target price is US$65/dz, if you can accept, please confirm and send one free sample for quality approval.

We are looking forward to receiving your reply soon.

Best regards,

3. 買方：寫一封還價信，告知賣方報價太高，無法與市場上的其他競爭者競爭，請考慮再降價 5% 我方才有興趣下單。

Dear Sirs,

Thank you for your quotation dated June 20. However, we regret to inform you that your price is too high to be accepted, which is difficult to compete in our market. Please kindly consider lowering the price 5%. This is the only way that we can consider placing an order with you.

We look forward to your acceptance and confirmation further.

Best regards,

4. 賣方：回覆客戶還價，表明已報最好的價格，無法再降價 5%，且買方的數量少產品多，實在沒有任何降價的空間，請對方了解。如果對方未來可以將數量增加到一定數量，我方可以再降價。

Dear Sirs,

We received with attention your counter offer on June 25. Please kindly note that we have quoted our best prices. It's indeed hard for us to lower 5% further. In addition, another problem is that you only order a small quantity for each product, which causes us an increase on production cost, although you order many different products. It's indeed no room for us to give you any discount further. Please kindly understand and try your best to increase your order quantity. For your future order in big quantity, we'll do reconsider about the price.

Best regards,

5. 賣方：為表示對客戶的支持，我方將於這一張訂單考慮接受客戶的目標價格：FOB Taiwan US$3.50/pc，但是訂單需於一星期內讓我方收到，並且我方只提供標準顏色。

Dear Sirs,

Your comments dated July 5 have been with our best attention. For supporting you, we can consider accepting your target price: FOB TAIWAN US$3.5/pc. However, please kindly let us receive your order in one week and accept our standard colors.

We hope our concession can result in your order confirmation and your acceptance of our conditions. We look forward to your favorable response soon.

Best regards,

6. 賣方：因手之日的原物料及工資上漲，我方將於明年度調漲價格，今年的價格將盡力維持原價，敬請及早下單，以舊價取得商品。

Dear Sirs,

Due to the increase of raw materials and the labor cost, we will be compelled to adjust prices next year. What we can do is trying our best to maintain our old prices by end of this year. Please kindly place your order as early as possible in order to get the old prices.

Thank you first for your kind attention, and we look forward to your order information soon.

Best regards,

Chapter 07

1. 將下列的句子翻成英文

A. We are pleased to receive your letter of May 12, and advise you of our sending the samples you need.
B. We are sending you our catalogue and samples under separate cover.
C. As instructed, we are sending our latest catalog in which all of our products are illustrated and specifications arc attached.
D. Upon your request, we are sending the samples,catalogs and price list for our cell phone.
E. We are glad to send you under separate cover our quotation and our sample.
F. We will place a sample order with you.
G. Before placing the order, we hope to receive your sample.
H. As requested, we are sending you one free sample by APP for quality approval.
I. We hope to receive your sample order very soon.

2. 將下列中文書信翻譯成英文

Dear Sirs,

We confirm the acceptance of your prices quoted in your quotation of Q2017158. However, before placing the order, please send us 2 free samples for evaluation. We will place you our formal order upon sample approval.

We hope to receive the above samples within one week, please speed them up.

Best regards,

3. 賣方：就以上來函，回一封信。

Dear Sirs,

Thanks for your email yesterday confirming your acceptance of our prices and requesting our samples. We are glad to inform you that the samples will be sent tomorrow by express air parcel post for your quality approval.

We are awaiting your order soon.

Best regards,

4. 請寫一封索取手錶樣品的索樣信函。

Dear Sirs,

I have seen your website and am interested in your range of watches, especially model No. AA-125. Could you quote these series of watches and send me a free sample of this item for evaluation?

Best regards,

5. 根據第 4 題的信函回覆客戶，因手錶價格高，因此要索取樣品費 US$150，但若是客戶日後下單，樣品費會從訂單中扣除。

Dear Sirs,

Thank you for your email in which you asked for a quotation for our new watch series and a free sample of item AA-125. We are pleased to quote this series watches and we are sure of meeting your market demand through your powerful distribution channels. Attached please find our best quotation based on FOB Taiwan.

However, concerning the sample you required, please kindly pay us US$150 which includes the sample cost and express shipping cost. As this item is expensive, please kindly understand, it's our company's policy for treating sample case. When you place an order in the future, we will deduct this charge from the order payment.

Your understanding and cooperation will be appreciated. We look forward to receiving your confirmation soon.

Best regards,

6. 寫一封信函通知客戶對其索求的樣品可以免費提供，但是客戶需付運費

Dear Sirs,

Thank you for your email of October 11 concerning the samples. We will prepare the samples soon and send them before October 15.

For the samples you requested the total amount is about US$67. For supporting your business, we offer these samples free of charge. However, the courier cost is very high and must be paid by your company. Please kindly inform us of your regular carrier and your account number, so that we can send the samples "freight collect."

Thank you for your attention and we look forward to receiving your reply soon.

Best regards,

7. 樣品寄出一週後，請寫一封 e-mail，跟催客戶。

Dear Sirs,

It has been a one week since we sent the samples to you. We would like to know if you received them. The quotation for these samples was also included. If you have any questions, please don't hesitate to contact us.

We hope that these products will help you to extend your market and help your exhibition to achieve a successful result. We are here to give you the best support and look forward to good news from you soon.

Best regards,

Chapter 08

1. 將下列的句子翻成英文

A. I have just returned from the Show this morning and found that you have not responded to our emails sent while I as absent.
B. A sufficient amount of time has elapsed and we would appreciate any response from you.
C. With reference to our records, it has been a long time we have not contacted with you. We sincerely wish you having a prosperous business.
D. We have sent you our latest catalog and the best quotation for our products on September 1. We are wondering if you have well received them since we have received no any response from you till now.
E. In keeping the policy of offering our new developed products to the respective customer like you continually, we believe that you should have received our new leaflet sent on May 1.
F. Please kindly note that we have gotten a lot of orders recently. It is better for you to confirm order as soon as possible, so that we can reserve a space for you.
G. Due to the heavy demand of this product, we would advise you to place your order as soon as possible.
H. Though prices have increased steadily since February, we have still tried our best to maintain our quotation. We hope that you will let us have your order before further rises in costs make an increase unavoidable.
I. As our production schedule is very tight, we would suggest you place order by the end of this month.
J. We assure you that the quality is equal to the sample we sent you. Your prompt order confirmation will be appreciated.
K. Our new products have been selling extremely well and we would like to recommend them to you with confidence. We have recently been receiving a lot of orders, we would advise to place your order soon.

2. 將下列中文書信翻譯成英文

Dear Sirs,

We sent you our latest catalog of lady's fashion shoes on 4/15 and would like to know if you have received it.

These shoes are the most fashion styles this year and they are welcome by consumers. If you are interested in any style, please decide your order soon or write us for the sample, we are glad to offer 2 free samples for customer's evaluation.

We hope to hear your interest and comments soon.

Best regards,

3. 寫一封追蹤信，表示我方已於 6/18 寄給客戶報價與樣品，但是一直到今日尚未收到回音，想請問有沒有收到樣品，看看客戶有無任何意見。如樣品不符客戶需求，我們也可根據客戶的規格打樣，再送客戶所需的規格樣品。

Dear Sirs,

On 6/18, we sent you our best quotation and samples. We are wondering if you have well received them since we have not received any response from you till now. Do you have any comments on price or sample?

We can revise our spec. upon client's request. If our sample does not meet your requirement, we can make sample according to your specification and send the sample for your approval.

Best regards,

4. 寫一封對老客戶的追蹤信，表示距上次訂單已有一段時間，希望不是因為客戶不滿我方的服務。我們保證客戶在和我們交易後都會非常滿意，如果您未下單是因為處理的商品種類改變，請讓我們知道，看看是否可以滿足您的需求。希望您還是銷售同款運動商品，因此我們在隨函附上最新的型錄。我們認為這份型錄在產品系列、品質和價格上，比其他業者優異。您從目錄中可看到，隨著匯率管制以及官方措施的取消，與上次訂單相比，我們的交易條件優惠許多。

Dear Sirs,

We notice that it is some time since we last received an order from you. We hope this is in no way due to dissatisfaction with our service or with the quality of goods. In either case we would like to hear from you.

We are most anxious to ensure that customers obtain maximum satisfaction from their dealings with us. If lack of orders from you is due to changes in the type of goods you handle, we may still be able to meet your needs if you will let us know in what directions your policy has changed.

As we have not heard otherwise, we assume that you are still selling the same range of sports goods, so a copy of our latest illustrated catalog is enclosed. We feel this compares favorably in range, quality and price with the catalog that our terms are now much better than previously, following the withdrawal of exchange control and other official measures since we last did business together.

We are looking forward to hearing from you soon.

Best regards,

5. 寫一封參展回來後的追蹤信，表示在CES展覽後，感謝對方來訪問我們的攤位，問其看過目錄後的意見或是興趣，再次強調我們產品的特色，最後希望有機會建立商業關係。

Dear Sirs,

Thank you very much for your coming over to our Booth during the CES last month.

As you have seen the samples we displayed on the booth, you should have a clear idea about the excellent quality of our products and all the advantages are listed in our catalog. Please don't hesitate to let us know of your interests or comments.

Thanks for your consideration and hope to have the change to set up the business relationship with you very soon.

Best regards,

Chapter 09

1. 將下列的句子翻成英文

A. Due to high market demand, we would like to increase the order quantity for 6000pcs. Please revise the order quantity and confirm by return email.
B. We are sorry to ask you to reduce 1000pcs from PO# XXX due to economic depression.
C. We agree to accept your revised quantity, please modify the order quantity.
D. We are sorry we cannot accept your quantity increase because we have completed the production and it is difficult to prepare the material for the increased small quantity.
E. Please confirm your order at the price quoted.
F. If you find the terms acceptable, please confirm the order subject to the above-mentioned modifications.
G. Any order that you may place with us will have our prompt and careful attention.
H. Thank you for your quotation QQ123. We accept your price and terms and would like to confirm our order no. AAA as attached. The formal order will be followed by post.

2. 將下列中文書信翻譯成英文

Dear Sirs,

We accept your price and delivery time in your quotation, and are glad to enclose our new order no. AA-2017888. Please do ship the goods on time because we have to catch the Christmas sale.

Please confirm this order and send us P/I to us for opening L/C.

Best regards,

3. 賣方：請就以上的信函，回一封完整的確認函。

Dear Sirs,

Thank you for your order no AA-2017888 and enclose here is our P/I for opening L/C. After receiving your L/C, we will arrange the production immediately and confirm we will ship these goods on time.

We are waiting for your L/C soon.

Best regards,

4. 請寫一封拒絕接收訂單的信函，表示對客戶的訂單的歡迎，但是所定的商品本公司已不生產。我們有新的商品，品質更好，功能更佳，雖然價格會高一些，但是可以推薦給客戶作為替代品的考量。

Dear Sirs,

Thank you for your email dated July 25 in which you placed an order for our item no. AA.

We appreciate your interest but regret to inform you that we no longer produce this product. We put forward our new developed model BB for your consideration, which we feel is an excellent replacement.

The quality, function and specifications are greatly improved at very little extra cost and are enjoying excellent sales. Concerning the price, we offer you the same price as the older model if you can place an initial order exceeding USD8,000. If your order does not exceed this amount, the price will be an additional 3%. We have confidence in this new model and are sure that it will give you complete satisfaction.

We have sent you a sample for evaluation by Federal Express. We hope you will receive it by end of this week and will confirm an order with us in the near future.

Best regards,

5. 請寫一封拒絕接收訂單的信函，客戶下訂單但是付款方式為出貨後 60 天電匯，說明我們的付款條件為出貨前全額付清或是信用狀方式。

Dear Sirs,

Thank you for your order confirmation dated Oct. 5, but we have noticed that your terms of payment is by T/T 60 days after B/L date, which we are unable to comply based on our company's policy.

According our our policy, terms of payment is by T/T in advance (before shipment) or by an irrevocable L/C at sight in our favor. Please kindly reconsider and return your confirmation.

Your kind acceptance of payment terms will be highly appreciated. We look forward to doing business with you further.

Best regards,

Chapter 10

1. 將下列的句子翻成英文

A. With new customers, our policy to request the certified financial statements, credit references and the name of your bank.
B. In adherence with our policy, we ask you to provide credit references.
C. It is routine to ask for the latest financial statements when a customer places a large order.
D. You were listed a a credit reference for ABC company.
E. We respect your privacy and will never share any information you provide to us.
F. Financial information is always kept confidential.
G. We would appreciate your providing us with any information you might have.
H. You may be assured that we will hold this information free from any responsibility on your side.
I. Could you give us some information on their business activities and financial status?

2. 將下列中文書信翻譯成英文

Taiwan Bear International

3F, No. 11, Park Avenue II
Science-Based Industrial Park
Hsin-Chu 30075, Taiwan
Tel: 886-03-5798888
Fax: 886-035978891

February 19, 20xx

Arthur Teveli
Sales Dept.
Eco International
323 Bannock Street
Denver, CO 80204
USA

Dear Mr. Teveli,

A purchase order from Pioneer Corporation, US, has listed you as a credit reference.

We would appreciate any information you can provide about Pioneer Corporation's credit history with your company. Key information includes how long the company has had an account with you and whether Pioneer has any outstanding debts. Be assured, we will keep any information you send us confidential.

Thank you for your assistance in this matter. Enclosed please find an addressed, postage-paid envelope for your convenience.

Sincerely yours,

Linda Chen

Linda Chen
Sales Manager

Enc.

3. 請寫一封向客戶往來銀行徵信的信函。

Dear Sirs,

Your bank has been given to us as a reference by Jason & Son who wants to do business with us recently.

We should appreciate very much your giving us some information on their financial standing. Any information about this company from you will be kept confidential. Your support will help us in our decision.

Best regards,

4. 請寫一封向商業徵信所對新客戶的信用徵信的信函。

Dear Sirs,

We have received a first order worth $10,000 from Jason & Son, who has requested open account terms. We have no information about this company, but as there are prospects of further large orders we should like to meet this order and provide open account terms if it is safe to do so.

Please let us have a report on the reputation and financial standing of the company and whether it is advisable for us to grant credit for this first order. We would also appreciate any advice on the maximum amount for which it would be safe to grant credit on a quarterly account.

Thanks for your help in advance.

Sincerely yours,

5. 請寫一封向客戶所提供的備查人進行客戶信用徵信的信函。

Dear Sirs,

Jason & Son wish to open an account with us and have given your name as a reference.

Please let us have your view on the firm's general standing and your opinion on whether they will be able to settle their accounts promptly with a credit up to USD10,000. We will, of course, treat all information in strict confidence.

We enclose a stamped, addressed envelope for your reply.

Sincerely yours,

Chapter 11

1. 將下列的句子翻成英文

A. Please open an irrevocable letter of credit in our favor.
B. We shall be glad if you will open an L/C without delay so that we may ship your goods as contracted.
C. We express our thanks for your L/C in the amount us USD50,000.
D. Please amend the L/C.
E. Please extend the validity and shipping time of the L/C to August 31.
F. We request you either extend the period of shipment to July 31, or amend the credit so that we may transship the goods.

2. 將下列中文書信翻譯成英文

Dear Sirs,

Regarding your PO No. AA254, we are glad to inform you that the goods are ready, but we have not received your L/C yet.

Please advise if you have opened L/C. If yes, please advise L/C details or scan the copy to us.

Best regards,

3. 請寫一封跟催信用狀的信函。

Dear Sirs,

With regard to our P/I#5526 sent on Feb. 22 and your confirmation on Mar. 10, we haven't received the L/C from you yet. Would you please open it immediately or kindly inform us when you expect to send us the L/C?

We plan to complete your order and ship it around September 22. As you can see, there is no time left, your L/C is urgently expected. Please take our notice as a serious matter and give us your reply as soon as possible. Thank you very much.

Best regards,

4. 請寫一封出貨前跟催電匯的信函。

Dear Sirs,

We are pleased to inform you that your order will be completed and we estimate goods will be shipped around July 20 by both air and sea freight.

Herewith, we would like to request that you arrange the payment of USD5,000 for the balance of this order within this week, so that we can settle your account before shipment as agreed.

We are glad to have been of service and look forward to your remittance soon.

Best regards,

5. 請寫一封信函通知客戶修改信用狀裝船期自 8/10 延至 8/30，有效期自 9/10 延至 9/30，以符合貨品出口日期，請客戶儘速修改並告知。

Dear Sirs,

Please refer to your PO#2933, L/C No. 928929199. We have just received your original L/C today for which we thank you very much. However, we found there are two terms should be amended.

1. Shipping date: Aug. 10, 20xx AMEND TO Aug. 30, 20xx

2. Expiry date: Sep. 10, 20xx AMEND TO Sep. 30, 20xx

When we received your L/C, it's already Aug. 3rd. We cannot ship the goods before Aug. 10, but in end of August. So, an amendment about the above is needed.

However, we'll try our best to ship it as early as possible and confirm this matter further. Please kindly amend them accordingly.

Thank you in advance, and we look forward to your L/C amendment soon.

Best regards,

6. 請寫一封信感謝客戶我們已收到信用狀，號碼 :880128，但是我們發現信用狀上有 3 點錯誤，(1) 價格條件應該是 FOB Twiwan(2) 應該要允許轉運 (3) 最後出貨日應為 10 月 30 日。以上請儘快修改並確認。

Dear Sirs,

We received your L/C no. 880128 with thanks, but we found following mistakes:

1. The price term should be FOB Taiwan.
2. Transshipment should be allowed.
3. The latest shipping date should be 10/30.

Please amend the above soon and confirm by return mail.

Best regards,

Chapter 12

1. 將下列的句子翻成英文

A. We have booked shipping space on the China ACE scheduled to set sail from Keelung around November 3.
B. Shipping space has been booked on the Hang Yang, which is scheduled to leave Kaohsiung on or around Nov. 3.
C. Your goods are nearly ready for dispatch, and we should be glad to have your immediately instruction.
D. We wish to confirm the dispatch of 100 sets of cameras by CAL Flight No.520 to New York.
E. Delivery will be made immediately on receipt of your L/C.
F. We are pleased to inform you that the consignment was collected this morning for road transport to Chicago.
G. We trust that your goods will reach you safely, and that you will favor us with further orders.
H. As requested, the triplicate commercial invoice and duplicate inspection certificate are sent to you.
I. We enclose packing list, bill of lading, invoice, certificate of origin and inspection report.
J. Your goods was airfreighted today with shipping documents airmailed.
K. The shipping documents will be sent to you by our forwarder.
L. The shipping documents listed below have been forwarded through Bank of Taiwan to Bank of America, New York with draft.
M. We cannot ship goods on the date you requested.
N. Due to port congestion toward the end of the year, it may not be possible to ship the entire order by the time of shipment.
O. The earliest delivery we can promise at present will be next January.
P. There is no ship to your port during August.

2. 將下列中文書信翻譯成英文

Dear Sirs,

Regarding PO no. 123, we are glad to inform you that we have shipped 100 cartons of woman sport shoes today via Ever Green Lines, s.s. "EVA V-456", ETD Keelung on 8/12 and ETA San Francisco on 8/20.

We have sent one set of non-negotiable documents today for your reference. We believe the goods will arrive your port in good condition and thanks for your patronage.

Best regards,

3. 請寫一封信告知客戶，貴公司訂單 PO123 已於今日裝船。船名為總統號，預定 8/8 由台中港駛向紐約港，到期日預計為 8/25。

Dear Sirs,

Shipping Notice

We would like to inform you that the shipping has been effected by vessel from Taichung port, Taiwan to New York. Shipping information as follows:

PO No.: 123

Delivery Date: 8/5
Vessel: Santa Maria V.0032
ETD: 8/8/20xx
ETA: 8/25/20xx

Best regards,

4. 請寫一封信告知客戶因地震的關係，我們無法以訂單原訂出貨日出貨，會延期出貨日至 10 月底，並請客戶修改信用狀之出貨日與到期日，一但確認出貨船期將立即告知。

Dear Sirs,

Due to a devastating earthquake, our factory suffers from serious damages. We regret to inform you that the delivery of your order No. 928 will be postponed by end of October.

It is why we have to ask a favor of you to amend L/C to November 15. We will inform you the shipping details immediately when available.

Best regards,

5. 請寫一封空運出貨的通知函，告知客戶已遵其指示，將貨物一半空運，一半海運。空運主要是把航班，主提單 (Master Air Waybill/MAWB) 及空運公司發的提單 (House Air Waybill/HAWB) 一併給客戶。海運會在確認船班後告知詳細資料。原始出貨文件將以 UPS 寄送。

Dear Sirs,

We have received with attention your instruction dated May 5 regarding your order no. PO-789456 which you asked us to ship by Sea and by Air.

As requested, we have delivered half of the goods to your airfreight forwarding agent. The flight details are listed below:

MAWB: 160HKG7684 4854
HAWB: HKG0173 1838
Airline: Cathy Pacific Airways CX-082 (June 1)

For the rest of the goods, we expect to ship them around July 8. We will confirm further about the vessel name, ETD and ETA. After the shipment is effected, we will send you the original documents by UPS.

Best regards,

Chapter 13

1. 將下列的句子翻成英文

A. Please remit the balance on the 15th of each month to our First Bank account.
B. Please be good enough to send us the amount due not later than the end of this month.
C. May we remind you that your payment has been overdue since May 10?
D. We insist upon full payment of your account; otherwise, we shall be forced to take legal steps.
E. We must insist on receiving payment by March 31; failing this, we shall be compelled to take legal action.
F. You will oblige us by instructing your bankers to cable remittance within a week.
G. We urge that you make this settlement without delay.
H. We understand how you now stand, but we can hardly overlook the fact that your payments have been delayed so frequently.
I. We have frequently reminded you of the outstanding amount but have received no reply or remittance from you.
J. Your failure to pay on time is in turn causing us financial embarrassment.
K. According to our records, you have not yet paid for order no. AA-123.
L. We regret that we will have to inform our lawyers if payment not made immediately.
M. We thank you in advance if your payment is already in process.

2. 將下列中文書信翻譯成英文

Dear Sirs,

We enclose herewith our Statement No. 56412 showing total amount US$13,365.78 due in this month.

Please kindly settle the above outstanding payment by T/T before end of this month. Thanks for your patronage and are awaiting your payment.

Best regards,

3. 根據下列中文情境，寫一封催款信

Dear Sirs,

Regarding your outstanding payment due this month, we enclose our Debit Note 88-123 covering total amount US$8,542.

Since the goods were shipped end of last month, we hope to receive the payment before end of this month. Please advise us by email upon remittance.

Best regards,

4. 根據第 3 題，寫一封第二次的催款信。

Dear Sirs,

Second reminder

Re: Debit Note 88-123
Balance due: US$8,452

We gratefully regret that you keep in silence, without replying our e-mail of September 15. Please kindly note that this amount is more than one month overdue. Your immediate action in settling it is required and appreciated. Please kindly arrange the payment by T/T and return your confirmation soon.

Best regards,

5. 請寫一封催款信告知客戶，我方已對客戶積欠 2 個月的貨款 US$17,500 提醒過至少 2 次，但是都未收到客戶的回應。請客戶須於立即清償帳款，否則本公司將交由公司的律師處理此事。

Dear Sirs,

Your balance of US$17,500 for order no. AA-123 is now more than 60 days overdue. We wrote to you twice requesting payment of your nonpayment; however, I am afraid we still have not heard from you. Without a prompt response from you, we will have to turn the account over to our lawyers.

Best regards,

6. 請寫一封催款信表示我們已於 3 月 10 日即 3 月 23 日去函提醒客戶 2 月份的對帳單欠款餘額 US$6,800.23 的事情，至今尚未收到任何回覆，令人感到失望。因為彼此過去幾年的合作關係良好，所以我們更感失望。在此情況下，除非您於 10 天內回覆，否則我們必須認真考慮進一步的行動，已取回欠款。

Dear Sirs,

We are disappointed not to have heard from you since our two letters of Mar 10 and March 23 reminding you of the balance US$6,800.23 still owing on our February statement.

This is all the more disappointing because of our past good relationship over many years. In the circumstances, unless we hear from you within 10 days we shall have to consider seriously the further steps we could take to obtain payment.

Best regards,

7. 請寫一封最後通牒的催款信，表示我方已於 7 月 15 日再次去函通知客戶逾期已久的積欠金額 US$1,534.03 的事情，很驚訝至今仍未收到回覆。過去雙方合作關係一直很好，即使如此，我們仍不允許帳款無限期不繳付。除非您在本月底前結清，或是提出令人滿意的解釋，否則我們將被迫交由我們的律師處理此事件。

Dear Sirs,

We are surprised that we have received no reply to the further letter we sent to you on July 15 regarding the long overdue payment of US$1,534.03.

Our relations in the past have always been good. Even so, we cannot allow this amount to remain unpaid indefinitely. Unless the amount due is paid or a satisfactory explanation received by the end of this month, we shall be reluctantly compelled to put this matter in the hands of our attorney.

Best regards,

Chapter 14

1. 將下列的句子翻成英文

A. We have received serious complaints in regard to your TV sets.
B. We have to complain that the quality of the received goods is worse than that of the sample.
C. We are sorry to inform you that we had numerous complaints concerning your goods. The greater part of them has been returned.
D. We would like to receive your explanation of this inferior quality and to know what you propose to do in the matter.
E. Your goods are much inferior in quality to your counter sample and slightly different in shade. From the survey report enclosed, you will readily admit that the goods are inferior in quality to the standard we request.
F. To prove the inferior quality, we have enclosed a survey report by Lloyd's Surveyor. You will readily admit the reasonableness of our claim.
G. Due to such inferior quality, we must claim a large allowance.
H. To prove our statement, we are enclosing one of the samples and a cutting of the material received yesterday.
I. We just received the documents and took delivery of the goods but found to our regret that two cartons of the goods were missing.
J. Upon checking the goods received, we have found that several items on your invoice were not included.
K. Your invoice does not agree with the quantity received, 5 cartons being short.
L. We are disappointed to find that the quality of the goods you supplied does not correspond with that of the samples submitted.
M. The goods are not in accordance with sample.

N. We are surprised that some of the goods have been damaged.
O. Two cases in the consignment were damaged on arrival. We have marked the consignment note accordingly.
P. The goods we ordered on 7/15 have arrived in damaged condition.
Q. On unpacking cartons, we found that half of the dishes were broken.
R. Owing to the defective packing, some of the goods were damaged to such an extent that we were compelled to dispose of them at much reduced prices.
S. The damage appears to have been mainly caused by the faulty packing of the goods.
T. Please look into the non-delivery of the 2000 cell phones which we ordered on 4/19.
U. This delay is causing us great inconvenience, as we have promised our customers early delivery.
V. More than a month has passed since receiving your shipping advice, yet we have not heard anything about shipment from the shipping company.
W. The goods arrived at our port 7 days later than scheduled.
X. We ask you to make us compensation corresponding to our loss.
Y. Please advise if you will take the goods back or let us sell them at a discount of 50%.
Z. We are prepared to retain these unsuitable goods, but only at a substantially reduced price.
AA. Due to the extreme interiority of the goods, our clients will refuse to accept these goods even you may make a big discount.
AB. We will submit this claim to arbitration.
AC. If the cargoes cannot be found within a few days, we will file our claim for the full settlement of them.
AD. Please investigate this matter and adjust it without delay.
AE. Your immediate attention to this matter would be appreciated.

2. 將下列中文書信翻譯成英文

Dear Sirs,

We regret to inform you of some serious quality problem with your delivery of 10 cartons of sandal shoes on Feb. 3. Those goods appeared to be moldy on surface.

The package containing these goods appeared to be in perfect condition and we accept these goods without question. When I unpacked the contents, I noticed the damaged goods. I can only assume that this was due to careless handling at some stage prior to packing.

I am attaching a list of the damaged goods and hope you will replace them soon. They have been kept at our warehouse now, please advise how to arrange these goods.

Best regards,

3. 請寫一封抱怨信，表示收到訂單號碼 PO-8541 的出貨，但是發現商品的尺寸與顏色不對，而且有些箱子有破損的情形，請儘快告知處理方法。

Dear Sirs,

We just received the shipment of PO-8541, but we found following problems:

1. The size and color are not correct.
2. Some cartons are broken.

Please advise your disposal soon.

Best regards,

4. 請就第 3 題回覆，請客戶寄回不良樣品，以便分析原因，避免雙方誤解，再告知解決的方法。

Dear Sirs,

We are sorry to hear that there are some problems happened on the shipment of PO-8541. However, to avoid the misunderstanding, please send back some defective samples for evaluation. After analyze the problem, we will advise you our disposal.

Best regards,

5. 請就第 3 題回覆，抱歉品質與包裝不良，願給 2% 折扣。

Dear Sirs,

We have received with attention your comments on June 20. First we sincerely apologize to cause inconvenience. As there are some difficulties on blending the exact colors you want, we regret that the result cannot reach your satisfaction. We also feel sorry for cartons broken during shipment. However, we did have done our best, and the quality of both products and cartons should be all right in next delivery.

In order to resolve this problem, we would like to offer you a discount of 2% as a compensation for this order. Please kindly understand our sincerity and accept it.

Best regards,

6. 請寫一封回函告知客戶，對於客戶抱怨的包裝與數量短缺問題，我們正在吊查，將於 3 天內回覆。

Dear Sirs,

Your comments dated Nov. 20 about packing and shortage have been noted. We are looking into this matter and will reply you within three days.

Best regards,

Chapter 15

1. 將下列的句子翻成英文

A. We are interested in acting as your exclusive agent in the Asia Pacific area.
B. If you could grant us the sole agency, we will make every effort to lead our business to complete success.
C. We shall act as the exclusive sales agent for you and receive a commission of 10% on all sales in the whole territory of Taiwan.

D. We thank you for your letter of April 17 offering us the sole agency for your products in Taiwan.
E. We are represented in your country.
F. We are fully represented.
G. Should the position change in the future, we shall be in touch with you.
H. Please kindly advise us your interest in cooperating with us and let us know your terms and conditions for a sole agent.
I. We agree to grant you as our sole agent for a trial period of one year in order to get business moving.

2. 將下列中文書信翻譯成英文

Dear Sirs,

We know from the advertisement that you are looking for an Agent and are interested to know your terms and conditions.

With many years' sales experience, we know very well the market here and have very good sales network. So, we have the confidence to promote very well your products here in Taiwan.

Thanks for your consideration in advance and are waiting for your reply soon.

Best regards,

3. 請就第 2 題回覆客戶，接受對方成為我們商品 PT-MEL 的代理。

Dear Sirs,

Thank you for your e-mail dated June 12 expressing your interest in being our sole agent in Taiwan. We are pleased to have this opportunity to cooperate with you for penetrating this market.

In consideration of the steady increase in the demand for our PT-MEL, we believe that there is an enormous demand in your market waiting for us to explore. We are pleased to have you to act as our sole agent and hope both of us have a great success after establishing this relationship.

However, before we reach to an agreement, please kindly inform us more about your plans and opinions, ex: your estimated annual quantity and turnover, the payment terms and the details of the contact. After obtaining the information as mentioned above, we'll draw up an agreement further for your study.

We look forward to hearing from you soon.

Best regards,

4. 請寫一封婉拒對方請求代理的信函。

Dear Sirs,

Thank you for your e-mail dated June 12 expressing your interest in being our sole agent in Turkey.

Herewith we thank you first for your interest in our products and your proposal of being our agent. However, we feel it would be better to discuss it after greater understanding and more actual business contact. We suggest you to place an order or present your purchase plan first. After cooperating with each other for a couple of years, we can discuss the exclusivity further.

Your kind understanding to the above is appreciated. We look forward to your further comments and inquiry soon.

Best regards,

Chapter 16

1. 將下列的句子翻成英文

A. I would like to convey my warm congratulations on your appointment to the Board of Electrical Industries Ltd.

B. On looking through the China Post this morning I came across your name in the New Years Honors List. I would like to add my congratulations to the many you will be receiving.

C. I was sorry to hear the news of XXX's passing.

D. It gives us great pleasure to announce that on April 1 we have entered into a close association with K&M Accountancy Co. Ltd.

E. We are pleased to announce the opening of our new branch in Taipei on Sep. 1.

F. Mr. and Mrs. Smith request the pleasure of XXX company to celebrate the wedding of their daughter Mary Smith.

2. 將下列中文書信翻譯成英文

Dear Sirs,

Invitation for Taipei Electronic Show

This is to inform you that our company will attend the Taipei Electronic Show held in Taipei World Trade Center during Oct. 8~15. Our booth number is A-016.

You are mostly welcome to come over here to take a look. After the show, we can also have a further business discussion.

Please don't hesitate to let us know your flight details when you fix your visiting schedule, so that we can arrange the people to pick you up at the airport.

We are looking forward to seeing you soon.

Best regards,

3. 寫一封祝賀的信函，祝賀同儕在公司任職已十週年。表示公司的成功與成長都是仰賴像他這樣的員工，並且感謝員工的貢獻讓公司在這個產業上屹立不搖。希望大家繼續共事，並於這個特別的週年慶獻上恭賀之意。

Dear John,

This month makes your 10th anniversary as a member of staff of XXX company. We would like to take this opportunity to thank you for these past 10 years of fine workmanship and company loyalty.

We know that the growth and success of our company is largely dependent on having strong and capable staff members like you. We also recognize the contributions you make in helping us maintain the position we enjoy in the industry.

We are hoping that you will remain with us for many years to come and would like to offer our congratulations on this special anniversary.

Best regards,

4. 請寫一封通知函，告知客戶本公司於今年 7 月 1 日起將搬遷至台北市中心，請告知新地址與聯絡方式，並請客戶修正資料，於 7 月 1 日後將所有信件寄到新住址。

Notice of Removal

Please be informed that our company is going to move to a new and modern premise in the center of Taipei on and after July 1, 20xx. New address and contact information are listed as below:

Address: F1. 12, No. 5, Li-hsing Road, Taipei, Taiwan
Phone No.: 886-2-23985565
FAX No: 886-2-23986688
E-mail: sales@tbi.com.tw

Please kindly update your record and send all your correspondence to our new office from July 1, 20xx.

Thanks and best regards,

Chapter 17

1. 請依報上的求才廣告，應徵銷售人員一職，撰寫一封求職信函。

請參考 課文的內容及自己的特色，自行撰寫

2. 依自己的背景、特色，撰寫一篇履歷表。

請參考 課文的內容及自己的特色，自行撰寫

MEMO

MEMO

國家圖書館出版品預行編目資料

現代商用英文：書信與應用/廖敏齡, 陳永文編著. –
二版. – 新北市：新文京開發出版股份有限公司,
2022.01
面；　公分

ISBN　978-986-430-809-5（平裝）

1.CST:商業書信　2.CST:商業英文　3.CST:商業應用文

493.6　　111000180

現代商用英文－書信與應用（第二版）（書號:H207e2）

編著者	廖敏齡　陳永文
出版者	新文京開發出版股份有限公司
地址	新北市中和區中山路二段 362 號 9 樓
電話	(02) 2244-8188（代表號）
FAX	(02) 2244-8189
郵撥	1958730-2
初版	西元 2018 年 07 月 20 日
二版	西元 2022 年 02 月 01 日

建議售價：520 元

法律顧問：蕭雄淋律師

ISBN　978-986-430-809-5

新文京開發出版股份有限公司

新世紀．新視野．新文京－精選教科書．考試用書．專業參考書